An Introduction to Formal Logic

Second Edition

Richard L. Epstein

Advanced Reasoning Forum

Copyright ©2020 by Richard L. Epstein.

All rights reserved. No part of this work may be reproduced, stored in a retrieval system, or transmitted, in any form or by any means, electronic, mechanical, photocopying, recording, or otherwise, without the prior written permission of the Advanced Reasoning Forum.

Names, characters, and incidents relating to any of the characters in this text are used fictitiously, and any resemblance to actual persons, living or dead, is entirely coincidental. *Honi soit qui mal y pense*.

The moral rights of the author have been asserted.

Advanced Reasoning Forum
P.O. Box 635
Socorro, NM 87801
www.AdvancedReasoningForum.org

ISBN   978-1-938421-52-5   paperback

ISBN   978-1-938421-53-2   e-book

An Introduction to Formal Logic

Preface to the Instructor

1 The Basics of Logic
    A. Propositions 1
    B. Propositions as Types 2
    C. Inferences 3
    Exercises 5

2 Reasoning with Compound Propositions
    A. Compound Propositions 7
    B. "Or" Propositions 7
    C. Contradictories 8
    D. Conditional Propositions 8
    Exercises 9
    E. Valid Forms of Inferences using Conditionals 10
    Exercises 12

3 Classical Propositional Logic: Form
    A. Propositional Logic and the Basic Connectives 13
    Exercises 15
    B. The Formal Language of Propositional Logic 15
    Exercises 18
    C. Realizations and Semi-Formal Languages 19
    Exercises 20

4 Classical Propositional Logic: Meaning
    A. Meaning 21
    Exercises 24
    B. Models 25
    C. Tautologies 26
    Exercises 28
    D. Semantic Consequence 29
    Exercises 32
    E. Semantically Equivalent Formulas 34
    Exercises 34

5 Using Classical Propositional Logic 36

6 Proofs
    A. Proving 43
    Exercises 46
    B. Deriving Consequences from Hypotheses 46
    Exercises 48

Summary of Chapters 1–6 50

7 Reasoning about Things 51

8 The Grammar of Things
- A. Names and Predicates 55
- B. Propositional Connectives 56
- C. Variables and Quantifiers 57
- D. The Grammar of Predicate Logic 59
- Exercises 59

9 A Formal Language for Predicate Logic
- A. The Formal Language 61
- B. Realizations: Semi-Formal English 63
- Exercises 66

10 A Predicate Applies to an Object or Objects
- A. Names 69
- B. Predications 70
- Exercises 74
- C. The Self-Reference Exclusion Principle 75
- D. The Universe for a Realization 76
- Exercises 77

11 Models for Classical Predicate Logic
- A. Assignments of References 79
- B. Valuations 81
- C. Compositionality and the Division of Form and Content . 81
- D. The Truth-Value of a Complex Proposition 82
- E. Models 85
- F. Tautologies and Semantic Consequence 88
- Exercises 89
- G. Propositional Logic in Predicate Logic 90
- H. The Division of Form and Content Verified 91

12 Substitution of Variables and Distribution of Quantifiers
- A. Order and Distribution of Quantifiers 93
- Exercises 96
- B. Substitution of Variables 97
- Exercises 100

13 An Axiom System for Classical Predicate Logic . 102

Summary of Chapters 7–13 105

14 Formalizing in Classical Predicate Logic
- A. Rewriting English Sentences 107
- B. Initial Criteria 107
- C. Common Nouns
  - Relative quantification with "all" 108
  - Relative quantification with "some" 109
  - Common nouns and multiple quantifiers 110
- D. Negations 113
- Exercises 115

    E. Categorematic Words and Meaning Axioms 115
    F. Adjectives and Adverbs 117
    Exercises 119
    G. Mass Terms 120
    H. Time 121
    I. Examples of Formalizing 123
    J. Criteria and Conventions of Formalization 131
    Exercises 135

15 Identity
    A. Identity 138
    B. Classical Predicate Logic with Equality 139
    Exercises 142
    C. Implicit Identity vs. Explicit Identity 142

16 Formalizing with the Equality Predicate
    A. Formalizing Other Quantifiers
        There are at least n 144
        There are at most n 145
        There are at exactly n 146
        "no" and "nothing" 146
        Quantifications we can't formalize 147
    B. Examples of Formalizing 148
    Exercises 153

17 Possibilities 156

Appendices

1 Proof by Induction 162

2 Set-Theory Notation 164

3 Naming, Pointing, and What There Is
    Agreements 165
    Naming, pointing, and descriptions 166
    Forms of pointing: what there is 168

4 Completeness Proofs
    Classical Propositional Logic (PC) 170
    Classical Predicate Logic 174
    Classical Predicate Logic with Equality 180

5 Other Interpretations of the Quantifiers and Variables
    The substitutional interpretation 182
    Naming all elements of the universe at once 183
    Surveying all interpretations of the name symbols . . . 184

6 Mathematical Semantics
    The Abstraction of Classical Propositional Logic Models 185
    The Extension of a Predicate 186
    Mathematical Models of Classical Predicate Logic . . 189

7 Aristotelian Logic

The Tradition	191
Why These Forms?	192
Subject and Predicate	193
Some Classifications of Categorical Propositions	194
Exercises	195
Truth and the Square of Opposition	195
Exercises	197
Syllogisms	198
Exercises	199
Evaluating Syllogisms for Validity	200
Exercises	202

Index of Symbols	203
Index of Examples	204
Index	211

Answers to the exercises can be found at:

    www.AdvancedReasoningForum.org/intro_formal_logic

Dedicated to my brother

Robert S. Epstein

whose help and encouragement made this book possible.

I am grateful to Juan Francisco Rizzo, Victoria Pöhls, Walter Carnielli, William S. Robinson, and Esperanza Buitrago-Díaz, who suggested many useful improvements. And I am especially grateful to Arnold Mazzotti, who worked through a draft of this book with me and whose help very much improved this text.

Preface to the Instructor

Logic is a tool we use to investigate how our language connects to the world so that we can reason better and, we hope, understand the world better. This book tells that story. It has a beginning, a middle, and an end.

We begin by setting out what formal logic is: the study of inferences for validity based on their form. Classical propositional logic is then presented as the simplest formal logic. In the development of that logic, the most important tools of formal logic are presented: a formal language, realizations, models, formal semantic consequence, and an axiom system. Examples of formalizing ordinary-language propositions and inferences show how to use classical propositional logic and also show some of its limitations.

The middle of the book relates form to the world. Predicate logic is motivated as a way to widen the scope of classical propositional logic to investigate more kinds of inferences whose validity depends on form. The large assumption that the world is made up of individual things is the basis for both the syntax and semantics of predicate logic. Besides form and the truth or falsity of atomic propositions, only the idea of assigning reference to terms as a kind of naming is needed. Many examples of formalizing show both the scope and limitations of classical predicate logic. Those depend on establishing criteria for what counts as a good formalization. The emphasis in those discussions is how the assumption that the world is made up of individual things is both useful and limiting.

The final chapter reflects on the success of the formal methods we've developed as means to reason to truths. Our formal logics circumscribe what we mean by "individual thing", namely, what can be reasoned about in predicate logic. Our formal logics give us a way to be precise about how we understand possibilities, though only relative to the assumptions we make about form and meaning and what there is in the world.

This story gives the basics, the fundamentals of formal logic. Along the way I point out how the work here can be extended and modified to apply to a wider scope of what we can formalize from ordinary-language reasoning. The story is not finished. Not here, not elsewhere.

    *   *   *   *   *   *   *

Some Points about the Organization and Content

- Appendices
    The appendices contain material that is either more technical than many students want, or too philosophical for many students, or is supplementary to the main line of the story.

- The form of atomic wffs
    Rather than take "Ralph is a dog" as a wff, we separate the roles of names and predicates. Thus, we write "(— is a dog) (Ralph)". We have a choice between "(— lives in —) (Arf, New Mexico)" or "(— lives in New Mexico) (Arf)" depending on whether we take "New Mexico" to be a name of a thing. This leads the student to see more clearly the roles of names and predicates and is

crucially important in extending classical predicate logic to allow for formalizing reasoning that involves relative adjectives and adverbs in *The Internal Structure of Predicates and Names*.

- Superfluous quantification in the formal language of predicate logic
    In most logic texts, the definition of a formal language for predicate logic allows for superfluous quantifications. The rationale for including such formulas is that it simplifies the definition of the formal language, allowing a definition of bound and free variables to be made later. But the disadvantage is that a formula such as "$\forall x \,((-\text{ is a dog})(\text{Ralph}))$" that would correspond to the nonsensical "For everything, Ralph is a dog" is deemed acceptable. The semantics for superfluous quantifiers treat that formula as equivalent to "Ralph is a dog", which can be true. That is not consonant with our normally treating nonsense as false in our reasoning, as you can see in "Truth and Reasoning" in my *Reasoning and Formal Logic*. The advantages of not allowing superfluous quantification, beyond ridding our semi-formal languages of nonsense, are significant: we need no axiom schemes for superfluous quantification, and many proofs about the language are simplified by no longer having to treat cases of superfluous quantification separately.

- Proof theory
    Hilbert-style axiomatizations of classical propositional logic, classical predicate logic, and classical predicate logic with equality are presented in the text, and their completeness proofs appear in an appendix. Natural deduction and other methods of proof can be left until those skills are needed.

- Functions
    Only in some parts of mathematics and science are functions used, and then partial functions are essential. To extend classical predicate logic to allow for formalizing reasoning with partial functions we would need to analyze how to reason with non-referring names, descriptive names, and descriptive functions. I do that in *The Internal Structure of Predicates and Names*.

- The History of Logic
    It's a big subject. It's complicated. And it's not illuminating at this level. I present a lot in my books *Propositional Logics*, *Predicate Logic*, *Computability*, and *Classical Mathematical Logic*.

- English and Formal Logic
    Some might object that the development of formal logic here is too closely tied to motivations and examples from English. But in slightly modified form the examples and motivation here will apply to reasoning in many other languages. If they do not serve, then that would be evidence that the notion of thing is not as deeply embedded in the other language, as I explain in *Predicate Logic* and "Nouns and Verbs" in *Language and the World: Essays New and Old*.

1 The Basics of Logic

    A. Propositions 1
    B. Propositions as Types 2
    C. Inferences 3
    Exercises 5

We want to know what to believe. What counts as good reason to believe? How can we codify what we know? What are the consequences of assuming this rather than that? Logic helps us answer these questions. It's a tool to help us reason well.

A. Propositions

We want rules to help us find what's true. But what is it that's true or false? Not apples or cars or people, except in a metaphorical sense. What we say is what is true or false.

Example 1: Suzy: *Spot is out of the yard.*
Analysis This is true or false.

Example 2: Dick: *No cat can swim.*
Analysis This is true or false.

Example 3: Zoe (to Dick): *Spot is chewing your shoe.*
Analysis This is true or false.

Example 4: Dick: *Bad dog, Spot, bad dog!*
Analysis This isn't true or false; it's just Dick trying to influence Spot, like a command.

Example 5: Zoe: *What time does the movie start?*
Analysis This isn't true or false.

Example 6: Dick: *Get me a beer, Zoe.*
Analysis This isn't true or false.

    Questions, commands, and a lot more that we say isn't true or false. But it's not just the words we say. It's the way we use them.

Example 7: Maria: *I wish I could get a job.*
Analysis Maria has been trying to get a job for three weeks and said this to herself late at night. It isn't true or false. It's more like a prayer or an extended sigh.

Example 8: Dick: *I wish I could get a job.*
Analysis Dick's parents have been berating him for not getting a job and he told them that it's not that he's not trying. Then he said this, which is true or false.

Proposition A *proposition* is a written or uttered sentence used in such a way that it is true or false, but not both.

We call truth and falsity the *truth-values* of propositions. In what follows, I'll just say "uttered" when I mean written or uttered.

We don't have to make a judgment about whether a sentence is true or whether it's false in order to classify it as a proposition. We need only judge that in the context in which the sentence is uttered, it's one or the other.

Example 9: Zoe: *Dick is upset.*
Analysis Zoe said this in a whisper to Tom when she and Dick were visiting him. Tom might not know whether it's true or false, but he knows it's one or the other. It's a proposition.

Example 10 : $2 + 2 = 5$
Analysis Mathematical formulas are meant to be true or false, so this is a proposition, a false one. A proposition need not be a sentence put forward as true.

B. Propositions as Types

We want to reason together. When I write "Spot is a dog", you understand the this as a proposition. If you then write "Spot is a DOG", or someone else writes "Spot is a dog", or a friend shouts "Spot is a dog", we understand those as the same proposition. They are the same words in the same order.

Words are Types In the course of any reasoning, any word we use will continue to have the same properties when used again. We agree to identify the two uses as the same word. Briefly, *a word is a type*.

Propositions are Types In the course of any reasoning, we will consider a sentence to be a proposition only if any other sentence or phrase that is composed of the same words and punctuation in the same order can be assumed to have the same properties of concern to us during that discussion. We agree to identify such sentences or phrases and treat them as the same. Briefly, *a proposition is a type*.

Example 11: Flo: *Spot is a dog.*
　　　　Berta: *Spot is not a dog. He's a cat.*
Analysis It seems that Flo and Berta disagree. But Flo is talking about Dick and Zoe's dog Spot, and Berta is talking about Mrs. Zerba's cat Spot. They are not using the word "Spot" the same.

Example 12: *Spot es un perro.*
Analysis Dick and Pancho are going back and forth between English and Spanish in reasoning about dogs and cats. They agree that they'll treat this sentence as being the same as "Spot is a dog."

Example 13: Rose rose and picked a rose.
Analysis We can't use this in our reasoning unless we distinguish the three inscriptions, using perhaps "$Rose_1$ $rose_2$ and picked a $rose_3$" or "$Rose_{name}$ $rose_{verb}$ and picked a $rose_{noun}$".

Example 14: I am 1.80 m tall.
Analysis This is true if Dick says it. It's false if Zoe says it.

We have to avoid words such as "I", "my", "now", or "this", whose meaning depends on the circumstances of their use. Such words are called *indexicals*, and they play an important role in reasoning. Yet our demand that words be types requires that they be replaced by words we can treat as uniform in meaning throughout a discussion, such as "Dick" for "I", or "March 9, 1991" for "now".

Example 15: It's raining.
Analysis Dick said this to Zoe today. It's true. Last Thursday, Zoe was outside when it was raining, then she went inside to answer the phone and told her mother, "It's raining" though it had actually stopped a moment before. That's false. So it seems a proposition can be true at one time and false at another. But it's incorrect to identify the two utterances of "It's raining" as being the same proposition. They're the same words in the same order, but they're different for our reasoning because one is true and one is false.

The device I've been using of putting quotation marks around a word or phrase is a way of naming that word or phrase, or any piece of language. We need some convention because sometimes it's not clear whether we're talking about a word or phrase. For example, if a professional tennis player tells Dick, "Love means nothing in tennis", it's not clear whether she's telling Dick what the word "love" means in tennis or asserting that there's no room for sentiment in tennis. When we talk about and mark off a piece of language with quotation marks, we've *mentioned* it. Otherwise, we *use* the word or phrase, as we normally do. Sometimes we use italics rather than quotation marks.

Sometimes people use quotation marks as the equivalent of a wink or a nod in conversation, a nudge in the ribs indicating that they're not to be taken literally or that they don't really subscribe to what they're saying. Used that way they're called *scare quotes*, and they allow us to get away with "murder".

C. Inferences

We're concerned not only with which propositions are true, but which propositions follow from other propositions.

Example 16: All dogs bark, and Humberto barks. So Humberto is a dog.
Analysis Does "Humberto is a dog" follow from "All dogs bark" and "Humberto barks"? What does it mean for one proposition to follow from one or several other ones?

4 An Introduction to Formal Logic

Inference An *inference* is a collection of two or more propositions, one of which is the *conclusion* and the others the *premises*, that is intended by the person who sets it out as either showing that the conclusion follows from the premises or investigating whether that is the case.

When does an inference show that the conclusion follows from the premises? That depends in part on what kind of reasoning we are analyzing. Different conditions apply depending on whether we are concerned with arguments, explanations, mathematical reasoning, reasoning about cause and effect, or reasoning with prescriptive propositions, as you can read in my series of books *Essays on Logic as the Art of Reasoning Well*. However, for all kinds of reasoning, a fundamental criterion for whether the conclusion follows is that the inference is valid or strong.

Valid and strong inferences An inference is *valid* means that there is no way the premises could be true and the conclusion false at the same time.

An inference is *strong* means that there is a way for the premises to be true and the conclusion false, but all such ways are unlikely.

An invalid inference that is not strong is *weak*.

If an inference is valid or strong, then the conclusion *follows from* the premises; the conclusion is a *consequence* of the premises.

So Example 16 is a weak inference: there are lots of ways the premises could be true and conclusion false: Humberto could be a seal, or a parrot that's learned how to bark, or a philosophy professor who thinks he is a dog.

Example 17: *All dogs bark, and Humberto is a dog. So Humberto barks.*
Analysis This is valid. There's no way the premises could be true and conclusion false. I can't explain to you why that's so—either you recognize it or you don't.

But just because a proposition follows from some others doesn't mean that the reasoning is good. Example 17 shouldn't give you reason to believe that Humberto barks, because the first premise is false: some dogs have had their vocal cords cut, and others are ridgebacks that can't bark. An inference that is meant to convince someone, possibly yourself, that its conclusion is true is called an *argument*.

Example 18: *Dogs bark. So Ralph is a dog or Ralph is not a dog.*
Analysis This inference is valid because there is no way that the conclusion could be false. It's a tautology.

Tautology A *tautology* is a proposition if there is no way it could be false.

Exercises

1. What is a proposition?
2. Which of the following are propositions? If necessary, supply a context.
   a. Ralph is a dog.
   b. I am 2 meters tall.
   c. "The Queen, my lord, is dead", said by Seton to Macbeth in *Macbeth*.
   d. Feed Ralph.
   e. Did you feed Ralph?
   f. Strike three!
   g. Ralph believes that George is a goose.
   h. Ralph didn't see George.
   i. Whenever Juney barks, Ralph gets mad.
   j. If anyone should say that cats are nice, then he is confused.
   k. If Ralph should say that cats are nice, then he is confused.
   l. If Ralph should say that cats are nice, then Ralph is confused.
   m. There are an odd number of stars in the universe.
3. a. What does it mean to say that a word is a type?
   b. What does it mean to say that a proposition is a type?
   c. Should we identify the following as being the same proposition?
      "Ralph is a dog", said about my Ralph.
      "Ralph is a dog", said about Ralph Abernathy, whom I never met.
   d. Should we identify the following as being the same proposition?
      "All the dogs in the yard are barking", said by Zoe at 3 p.m.
      "All the dogs in the yard are barking", said by Dick at 8 p.m.
4. a. What is an indexical?
   b. Give two examples of an indexical used in a sentence.
   c. Explain why we choose not to allow indexicals to appear in the propositions we're considering here.
5. a. What is an inference?
   b. What is a valid inference? Give an example.
   c. What is a strong inference? Give an example.
   d. What does it mean to say that one proposition follows from one or more other propositions?
   e. What does it mean to say that one proposition is a consequence of one or more other propositions?
6. What is a tautology. Give two examples.

Aside: *Other conceptions of propositions*
Some say that what is true or false is not the sentence but the "meaning" or "thought" expressed by the sentence. Thus "Spot is a dog" is not a proposition; it expresses a proposition, the very same one expressed by "Spot is a domestic canine" and by "Spot es un perro".

Platonists take this one step further. A *platonist* is someone who believes that there are abstract objects not perceptible to our senses that exist independently of us. Such objects can be perceived by us only through our intellect. The independence and timeless existence of such objects account for objectivity in logic and mathematics. In particular, propositions are abstract objects, and a proposition is true or is false, though not both, independently of our even knowing of its existence. Thus each of the following, if uttered at the same time and place, expresses or stands for the same abstract proposition: "It is raining", "Pada deszcz", "Il pleut". Platonists say that the word "true" can be properly used only for things that cannot be seen, heard, or touched. Sentences are understood to "express" or "represent" or "participate in" such propositions. The assumption that propositions are types is said by them to be about which inscriptions and utterances represent or express or point to the same abstract proposition.

Those who take abstract propositions as the basis of logic argue that we cannot answer precisely the questions: What is a sentence? What constitutes a use of a sentence? When has one been been put forward for discussion? These questions, they say, can and should be avoided by taking things inflexible, rigid, and timeless as propositions. But then we have the no less difficult questions: How do we use logic? What is the relation of these theories of symbols to our arguments, discussions, and search for truth? How can we tell if this utterance is an instance of that abstract proposition?

In the end, though, the platonist as well as a person who thinks a proposition is the meaning of a sentence or a thought reasons in language, using sentences that they call "representatives" or "expressions" of propositions. We can and do reason together using those, and to that extent our definition of "proposition" can serve those folks, too.

Key Words	proposition	inference
	truth-value	conclusion
	types	premise
	indexical	valid inference
	quotation marks	strong inference
	use of a word or phrase	a proposition follows
	mention of a word or phrase	tautology
	scare quotes	

2   Compound Propositions

    A. Compound Propositions 7
    B. "Or" Propositions 7
    C. Contradictories 8
    D. Conditional Propositions 8
    Exercises 9
    E. Valid Forms of Inferences using Conditionals 10
    Exercises 12

A. Compound Propositions

Compound proposition   A *compound proposition* is one that has another proposition as part but has to be viewed as just one proposition.

Example 1:   Dick: Who won the election for mayor?
            Zoe: *Either a Democrat won the election or a Republican won the election.*
Analysis   This is a single proposition made up of two propositions "A Democrat won the election" and "A Republican won the election." They're joined by the word "or." Whether it's true depends on whether one or both of its parts are true. But the entire sentence is just one proposition.

Example 2:   *If Suzy studies hard, then Suzy will pass the exam.*
Analysis   This is just one proposition, made up of the two propositions "Suzy studies hard" and "Suzy will pass the exam."

Example 3:   *Lee will pass his exam because he studied so hard.*
Analysis   This is not a compound proposition: "because" tells us that this meant as an inference.

B. "Or" Propositions

Example 4:   *Dick or Zoe will go to the grocery to get eggs.*
Analysis   We can view this as an "or" proposition compounded from "Dick will go to the grocery to get eggs" and "Zoe will go to the grocery to get eggs." The propositions that make up an "or" proposition are called the *alternatives*.

Example 5:   *Either Dick picked up Zoe at the market, or Zoe went to see Suzy. Zoe didn't go to see Suzy. So Dick picked up Zoe at the market.*
Analysis   This is a valid inference. We don't need to know anything about Dick or Zoe or Suzy or the market to see that. It's valid just because of its form.

        Excluding Possibilities
         $\dfrac{A \text{ or } B \; + \; \text{not } A}{\downarrow}$ Valid
             B

I've used the letters A, B, and C to stand for any propositions, the symbol "+" to indicate that we have an additional premise, and ↓ to stand for "therefore".

C. Contradictories

Example 6: *Humberto got out of jail or he didn't get parole. Humberto did get parole. So Humberto got out of jail.*
Analysis This is an example of excluding possibilities. Here B is "Humberto didn't get parole" and the not-version of that is "Humberto did get parole." We can read "not-A" in excluding possibilities to mean the contradictory of A.

Contradictory of a proposition A *contradictory* of a proposition is one that must have the opposite truth-value.

Example 7: *Spot is barking.*
Analysis A contradictory of this is "Spot is not barking."

Example 8: *Inflation will be at least 3% this year.*
Analysis A contradictory of this is "Inflation will be less than 3% this year", which doesn't contain "not".

Contradictory of an "or" proposition A or B has contradictory *not A and not B*.
Contradictory of an "and" proposition A and B has contradictory *not A or not B*.

Example 9: *Maria got the van or Manuel won't go to school.*
Analysis A contradictory is "Maria didn't get the van, and Manuel will go to school."

Example 10: *Tom or Suzy will pick up Manuel for class today.*
Analysis A contradictory is "Neither Tom nor Suzy will pick up Manuel for class today."

D. Conditional Propositions

Conditional propositions A *conditional proposition* is a proposition that is or can be rewritten as one in the form *If A then B* that must have the same truth-value. The proposition A is the *antecedent* and B is the *consequent*.

Example 11: *If Spot ran away, then the gate was left open.*
Analysis This is a conditional with antecedent "Spot ran away" and consequent "The gate was left open." The consequent need not happen later.

Example 12: *I'll never talk to you again if you don't apologize.*
Analysis This is a conditional with antecedent "You don't apologize" and consequent "I'll never talk to you again."

Example 13: *Loving someone means you never throw dishes at him.*

Analysis This is a conditional with antecedent "You love someone" and consequent "You never throw dishes at him". It's not a definition.

Example 14: *A mammal is an ungulate if it has hoofs.*
Analysis This is not a conditional or a compound. It's a definition that uses "if" instead of "means that". We have to use our judgment to decide whether a proposition is a conditional.

Example 15: *If Dick goes to the basketball game, then either he got a free ticket or he borrowed money for one.*
Analysis This is a conditional whose consequent is a compound proposition.

Example 16: *Dick will go into the army only if there is a draft.*
Analysis What does "only if" mean? It means that if there is no draft, then Dick won't go into the army. And that is true just in case "If Dick goes into the army, then there is a draft". Generally, "A only if B" is true just in case "If A then B" is true. "Only if" does not mean the same as "if".

Example 17: *Dick will go into the army if and only if there is a draft.*
Analysis This is how we say "Both if Dick goes into the army then there is a draft, and if there is a draft, then Dick will go into the army".

"only if" propositions *A only if B* is equivalent to *If A, then B.*

Equivalent propositions Two propositions are *equivalent* means that the one is true exactly when the other is true.

"if and only if" propositions *A if and only if B* means *If A, then B, and if B then A.* We abbreviate "if and only if" as as "iff".

A contradictory of a conditional is not another conditional.

Contradictory of a conditional *If A, then B* has contradictory *A but not B.*

Example 18: *If Spot barks, then Suzy's cat will run away.*
Analysis Contradictory: Spot barked, but Suzy's cat did not run away.

Example 19: *If Spot got out of the yard, he was chasing a squirrel.*
Analysis Contradictory: Spot got out of the yard, but he wasn't chasing a squirrel.

Exercises
For each of the following: a. Say if it's a compound proposition. b. If it's a conditional, state the antecedent and consequent. c. Write a contradictory of it.
1. Maria or Lee will pick up Manuel after classes.
2. Neither Maria nor Lee has a bicycle.
3. AIDS cannot be contracted by touching or by breathing air in the same room as a person infected with AIDS.

4. Zoe (to Dick): Will you take the trash out, or do I have to?
5. If Spot barks, then Puff will run away.
6. Lee will take care of Spot next weekend if Dick will help him with his English exam.
7. Since 2 times 2 is 4, and 2 times 4 is 8, I should be ahead $8, not $7.
8. If Manuel went to the basketball game, then he either got a ride with Maria or he left early in his wheelchair to get there.
9. Drop the gun and no one will get hurt.
10. Maria will get a raise at work if and only if she is on time for work for a month.

E. Valid Forms of Inferences using Conditionals

Let's look at some valid forms of reasoning using conditionals and some forms that look a like those but are usually weak. Here is an illustration for some of the examples.

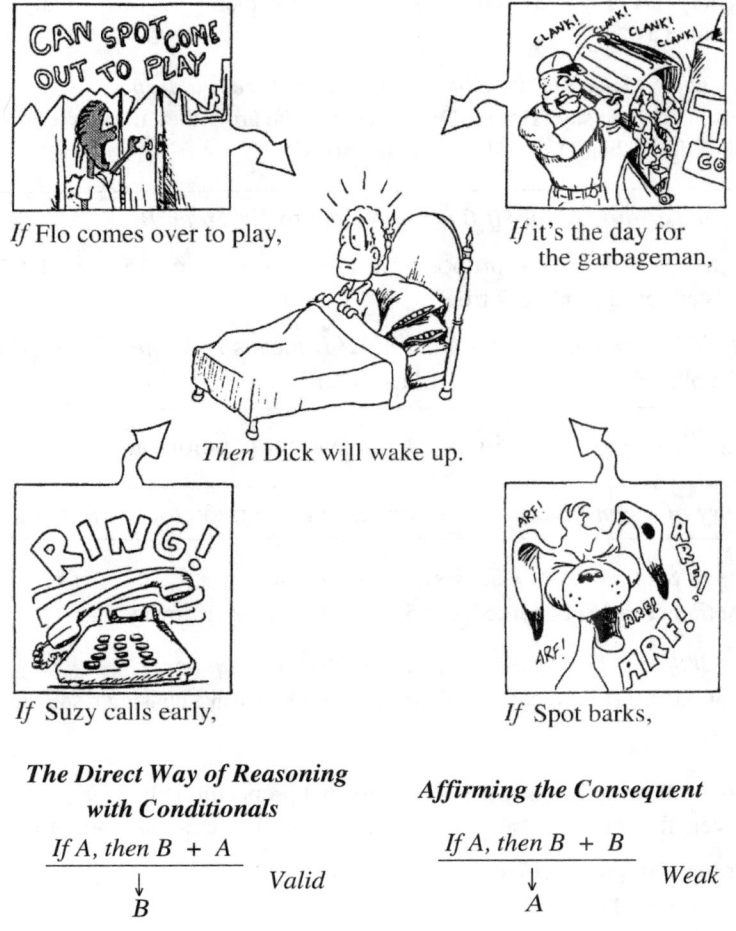

If Flo comes over to play,   *If* it's the day for the garbageman,

Then Dick will wake up.

If Suzy calls early,   *If* Spot barks,

The Direct Way of Reasoning with Conditionals

If A, then B + A
↓
B   Valid

Affirming the Consequent

If A, then B + B
↓
A   Weak

The direct way of reasoning with conditionals is often called *modus ponens*.

Example 20: *If Spot barks, then Dick will wake up. Spot barked. So Dick woke up.*
Analysis This is a valid inference. It is impossible for the premises to be true and conclusion false at the same time. It's an example of the direct way of reasoning with conditionals.

Example 21: *If Spot barks, then Dick will wake up. Dick woke up. So Spot barked.*
Analysis This is weak. Maybe Suzy called, or Flo came over to play. It's affirming the consequent, reasoning backwards.

The Indirect Way of Reasoning with Conditionals

$$\frac{\text{If } A, \text{ then } B \ + \ \text{not } B}{\text{not } A} \quad \text{Valid}$$

Denying the Antecedent

$$\frac{\text{If } A, \text{ then } B \ + \ \text{not } A}{\text{not } B} \quad \text{Weak}$$

The indirect way of reasoning with conditionals is often called *modus tollens*.

Example 22: *If Spot barks, then Dick will wake up. Dick didn't wake up. So Spot didn't bark.*
Analysis This is valid, an example of the indirect way of reasoning with conditionals.

Example 23: *If it's the day for the garbageman, then Dick will wake up. It's not the day for the garbageman. So Dick didn't wake up.*
Analysis This is weak. Even though the garbageman didn't come, maybe Spot barked or Suzy called early. It overlooks other possible ways the premise could be true and conclusion false.

Example 24: *If Maria doesn't call Manuel, then Manuel will miss his class. Maria did call Manuel. So Manuel didn't miss his class.*
Analysis This is weak, denying the antecedent. Remember that "not" in the form indicates a contradictory.

Reasoning in a Chain with Conditionals

$$\frac{\text{If } A, \text{ then } B \ + \ \text{If } B, \text{ then } C}{\text{If } A, \text{ then } C} \quad \text{Valid}$$

Example 25: *If Dick takes Spot for a walk, then Zoe will cook dinner. And if Zoe cooks dinner, then Dick will do the dishes. So if Dick takes Spot for a walk, then he'll do the dishes. But Dick did take Spot for a walk. So he must have done the dishes.*
Analysis This is a valid inference: reasoning in a chain with conditionals followed by the direct way of reasoning with conditionals. We conclude the last consequent because we have the first antecedent.

Reasoning from Hypotheses

If you start with an hypothesis A and make a valid inference with conclusion B, then you've shown that *If A, then B* must be true

Example 26: Lee: *I'm thinking of majoring in biology.*
Maria: *That means you'll take summer school. Here's why: You're in your second year now. To finish in four years like you told me you need to, you'll have to take all the upper-division biology courses your last two years. And you can't take any of those until you've finished the three-semester calculus course. So you'll have to take calculus over the summer to finish in four years.*

Analysis Maria has not proved that Lee has to go to summer school. Rather, on the assumption (hypothesis) that Lee will major in biology, Lee will have to go to summer school. That is, Maria has proved "If Lee majors in biology, then he'll have to go to summer school."

Exercises

Evaluate the following as valid or weak.
Identify the form of the inference if it's one from this chapter.

1. Tom: Either you'll vote for the Republican or the Democratic candidate for president.
   Lee: No way I'll vote for the Democrat.
   Tom: So you'll vote for the Republican.

2. Dick: Somebody knocked over our neighbor's trash can last night. Either our neighbor hit it with her car again when she backed out, or a raccoon got into it, or Spot knocked it over.
   Zoe: Our neighbor didn't hit it with her car because she hasn't been out of her house since last Tuesday.
   Dick: It wasn't a raccoon because Spot didn't bark last night.
   Zoe: Spot! Bad dog! Stay out of the trash!

3. If Suzy breaks up with Tom, then she'll have to return his letter jacket. But there is no way she'll give up that jacket. So she won't break up with Tom.

4. Steve Pearce is a congressman who meets with his constituents regularly.
   If someone is a good congressman, he meets with his constituents regularly.
   So Rep. Pearce is a good congressman.

5. Dr. E (on an exam day): If students don't like me, they won't show up. But all of them showed up today. So they must really like me.

6. Maria: Lee will take care of Spot Tuesday if Dick will help him with his English paper.
   Manuel: (*later*) Dick didn't help Lee with his English paper, so I guess Lee didn't take care of Spot on Tuesday.

3 Classical Propositional Logic: Form

> A. Propositional Logic and the Basic Connectives 13
> Exercises 15
> B. The Formal Language of Propositional Logic 15
> Exercises 18
> C. Realizations and Semi-Formal Languages 19
> Exercises 20

A. Propositional Logic and the Basic Connectives

We've looked at some inferences that are valid due to their form, and we've seen some whose forms look a lot like the valid ones but are weak. As well, some proposition are true due to their form. For example, we don't need to know anything about Dick and what he likes in order to know that the following is true:

> Dick liked the movie or Dick didn't like the movie.

Should we go on, looking at one form after another to try to decide whether it guarantees validity or truth? Do we have any method beyond our intuition?

We can do better. We can try to make precise what forms we're looking at. Then we can say how we'll understand the forms so that we can have a general method for evaluating propositions and inferences.

Formal logic *Formal logic* is (i) the analysis of inferences for validity in terms of the structure of the propositions appearing in an inference, and (ii) the analysis of propositions for truth in terms of their structure.

Formal logic will be a tool we can use in the analysis of inferences and propositions. By itself it can't determine whether any particular reasoning is good. Nor can it help us evaluate strong inferences.

To begin, we'll follow up on what we did in the last chapter, looking at simple forms based on how we can combine propositions.

Propositional logic *Propositional logic* is formal logic where we ignore the internal structure of propositions except as they are built from other propositions in specified ways.

Let's start with four phrases for building up propositions from other propositions we saw in the last chapter: the *connectives* "and", "or", "not", and "if . . . then . . .". These will give us a basis to see the issues in classifying forms and understanding meanings to serve us when we later extend what we pay attention to in the *syntax* (forms) and *semantics* (meaning) of propositions.

We use these English connectives in many ways, some of which may be of no concern to us in our reasoning. We'd like to agree on how we'll understand them and how they contribute to reasoning. But if we continue to use those English words, we're likely to forget how we've focussed on just this or that meaning. So let's use some symbols.

symbol	what it will be an abstraction of
∧	and
∨	or
¬	it's not the case that
→	if . . . then . . .

So a sentence we might study is "Ralph is a dog ∧ dogs bark".

When we talk about words or symbols, we should use quotation marks, saying "∧" and "→" are symbols. But too many quotation marks make a page look like a series of hen tracks. So let's agree that a formal symbol can name itself when confusion seems unlikely. So I can say ∧ is a formal connective.

Here's some terminology we use with these formal connectives:

- The sentence formed by joining two sentences with ∧ is called a *conjunction*. Each of the original propositions is a *conjunct* that we *conjoin* with the other.

- The sentence formed by joining two sentences with ∨ is called a *disjunction*. Each of the original propositions is a *disjunct* that we *disjoin* with the other.

- The sentence formed by joining two sentences with → is called a *conditional*. The proposition on the left is called the *antecedent* and the one on the right is called the *consequent*.

- The sentence formed by putting ¬ in front of a proposition is called the *negation* of that proposition. It is a *negation*.

Example 1: *If Ralph is a dog, then Ralph barks or Ralph howls.*
Analysis It seems that we should replace this English sentence with:

(a)   Ralph is a dog → Ralph barks ∨ Ralph howls

But this is ambiguous, though the example isn't because of the comma. Rather than using commas, let's use parentheses to mark off phrases.

Ralph is a dog → (Ralph barks ∨ Ralph howls)

Example 2: *If George is a duck then Ralph is a dog and Dusty is a horse.*
Analysis The original is ambiguous. Should we replace it with:

George is a duck → (Ralph is a dog ∧ Dusty is a horse)
or
(George is a duck → Ralph is a dog) ∧ Dusty is a horse

Only by asking the person who said the example or guessing from context can we make a choice. But it's clear we need to make a choice.

Exercises
1. Classify each of the following as: a conjunction (specify its disjuncts); a disjunction (specify its disjuncts); a conditional (specify its antecedent and consequent); a negation (specify what it is a negation of), or none of these.
   a. Ralph is a dog ∧ dogs bark
   b. Ralph is a dog → dogs bark
   c. ¬ cats bark
   d. Cats bark ∨ dogs bark
   e. Ralph is a dog.
   f. Cats are mammals and dogs are mammals.
   g. ¬ cats bark → ¬ cats are dogs
   h. Cats aren't nice.
   i. It is possible that Ralph is a dog.
   j. Either Ralph is a dog or Ralph isn't a dog.
2. a. Give a sentence that is a negation of a conditional whose antecedent is a conjunction.
   b. Give a sentence that is a conjunction of disjunctions, each of whose disjuncts is either a negation or has no formal symbols in it.
3. Give an example of a use of "not" that can't be understood as "It's not the case that".

B. The Formal Language of Propositional Logic

We want to talk in general about propositions. For the simplest one, the ones in which no formal connective appears, let's use the symbols $p_0, p_1, \ldots$. These are the *propositional variables*. These will be the first level of our formal language

We get more complex propositions by combining two of these using one of ∧, ∨, → or preceding one by ¬.

Level 1 p_0 p_1 p_2 p_3 ...

Level 2 $p_0 \wedge p_1$ $p_7 \wedge p_1$ $p_{83} \wedge p_{12}$ $p_4 \vee p_1$ $p_0 \to p_1$ $\neg p_5$ ...

At the next level, we get more complex forms by combining two of any of the preceding levels using one of ∧, ∨, → or preceding one by ¬ :

Level 3 $(p_0 \wedge p_1) \vee p_5$ $\neg (p_7 \wedge p_1)$ $(p_4 \vee p_1) \wedge p_5$
$(p_{83} \wedge p_{12}) \to (p_0 \to p_1)$

At each level we use one more formal connective to build a propositional form.

How can we specify exactly what is in each level? How do we say, in general, that we can join any two forms from the second level with → ? Let's use some symbols, not part of the formal language we're establishing, to stand for any of the formulas, at any level: $A, B, C, A_0, A_1, A_2, \ldots$. And instead of talking of p_i and p_j and natural numbers, let's use p, q, and r, to stand for any of the propositional variables. These are *metavariables*. We can then define the forms that we will look at with an *inductive definition*: We state what is the lowest level, and then we say that if we already have forms of some levels, we can create a new form by combining those with one of the four connectives, taking us up one level. This will give us all the forms of the propositions we'll be considering, a completely formal "language".

The formal language of propositional logic $L(p_0, p_1, \ldots, \neg, \rightarrow, \wedge, \vee)$

Vocabulary   *propositional variables* p_0 p_1 $\ldots$
             connectives $\neg$ $\rightarrow$ $\wedge$ $\vee$

Punctuation   *parentheses*) (

Grammar (*well-formed formulas — wffs*)

Each of (p_0), (p_1), (p_2), (p_3), $\ldots$ is a wff of *length* 1.

If A is a wff of length n, then $(\neg A)$ is a wff of *length* $n+1$.

If A and B are wffs and the maximum of their lengths is n, then $(A \rightarrow B)$, $(A \wedge B)$, and $(A \vee B)$ are wffs of *length* $n+1$.

A concatenation of symbols of the vocabulary is a *wff* iff it is a wff of length n for some $n \geq 1$.

Wffs of length 1 are *atomic*; all others are called *compound*.

It may seem obvious that parentheses ensure that if a concatenation of symbols is a wff, then there's only one way to read it. But that needs a proof. It's a proof by *induction*; if you're not familiar with that method, you read Appendix 1.

Theorem 1   The unique readability of wffs
There is one and only one way to parse each wff of $L(\neg, \rightarrow, \wedge, \vee, p_0, p_1, \ldots)$.

Proof To each primitive symbol α of the formal language, assign an integer $\lambda(\alpha)$ according to the following chart:

$\neg$	$\rightarrow$	$\wedge$	$\vee$	p_i	(	)
0	0	0	0	0	-1	1

To the concatenation of symbols $\alpha_1 \alpha_2 \cdots \alpha_n$, assign the number:

$\lambda(\alpha_1) + \lambda(\alpha_2) + \cdots + \lambda(\alpha_n)$

We'll first show that for any wff A, $\lambda(A) = 0$, using induction on the number of occurrences of $\neg, \rightarrow, \wedge, \vee$ in A.

If there are no occurrences, then A is atomic, that is, for some $i \geq 0$, A is (p_i). Then $\lambda('(') = -1$, $\lambda(p_i) = 0$, and $\lambda(')') = 1$. Adding, we have $\lambda(A) = 0$.

Suppose the lemma is true for every wff that has fewer occurrences of these symbols than A does. Then there are 4 cases, which we can't yet assume are distinct: A arises as $(\neg B)$, $(B \rightarrow C)$, $(B \wedge C)$, or $(B \vee C)$. By induction $\lambda(B) = \lambda(C) = 0$, so in each case by adding, we have $\lambda(A) = 0$.

Now I'll show that if α is an initial segment of a wff, reading from the left, other than the entire wff itself, then $\lambda(\alpha) < 0$; and if α is a final segment reading from the left other than the entire wff itself, then $\lambda(\alpha) > 0$. So no proper initial or final segment of a wff is a wff. To establish this I will again use induction on the number of occurrences of connectives in the wff. I'll let you establish the base case for atomic wffs, where there are no (that is, zero) connectives.

Suppose now that the lemma is true for any wff that contains $\leq n$ occurrences of the connectives. If A contains $n + 1$ occurrences, then it must have (at least) one of the forms given in the definition of wffs. If A has the form $(B \wedge C)$, then an initial segment of A must have one of the following forms:

i. (
ii. (β where β is an initial segment of B
iii. (B
iv. (B $\wedge$
v. (B $\wedge$ γ where γ is an initial segment of C

For (ii), $\lambda(\text{`(`}) = -1$, and by induction $\lambda(\beta) < 0$, so $\lambda(\text{`(}\beta\text{'}) < 0$. I'll leave (i), (iii), (iv), and (v) to you. The other cases (for $\neg$, $\to$, and $\vee$) follow similarly, and I'll leave those and the proof for final segments to you.

Now to establish the theorem, we proceed through a number of cases by way of contradiction. Suppose we have a wff that could be read as both $(A \wedge B)$ and $(C \to D)$. Then $A \wedge B)$ must be the same as $C \to D)$. In that case either A is an initial part of C or C is an initial part of A. But then $\lambda(A) < 0$ or $\lambda(C) < 0$, which is a contradiction, as we proved above that $\lambda(A) = \lambda(C) = 0$. Hence, A is C. But then we have that $\wedge B)$ is the same as $\to D)$, which is a contradiction.

Suppose $(\neg A)$ could be parsed as $(C \to D)$. Then $(\neg A$ and $(C \to D$ must be the same. So D would be a final segment of A other than A itself, but then $\lambda(D) > 0$, which is a contradiction. The other cases are similar, and I'll leave them to you. ∎

Example 3: (p_1) has length 1

Example 4: $(\neg(p_1))$ has length 2

Example 5: $((\neg(p_1)) \wedge (p_2))$ has length 3

Example 6: $(((\neg(p_1)) \wedge (p_2)) \to (p_3))$ has length 4.

$$\underbrace{\underbrace{\underbrace{(((\underbrace{\neg(p_1)}_{1}) \wedge \underbrace{(p_2)}_{1})}_{2} \to \underbrace{(p_3)}_{1})}_{3}}_{4}$$

It's hard to read a long wff because of all the parentheses. Those are needed to ensure that there is just one way to read each wff. But informally we can delete some of them if we adopt some conventions:

- We can drop the outermost parentheses.
- We can drop the parentheses around the propositional variables.

18   An Introduction to Formal Logic

- We'll say that $\neg$ binds more strongly than $\wedge$, $\vee$, or $\rightarrow$.
- We'll say that $\wedge$ and $\vee$ bind equally strongly, but both more strongly than $\rightarrow$.

Example 7: $\neg p_1 \wedge p_2$ abbreviates $((\neg(p_1)) \wedge (p_2))$

Example 8: $(p_1 \wedge p_2) \vee p_3$ abbreviates $(((p_1) \wedge (p_2)) \vee (p_3))$

Example 9: $\neg p_1 \wedge p_2 \rightarrow p_3$ abbreviates $(((\neg(p_1)) \wedge (p_2)) \rightarrow (p_3))$

These are informal abbreviations to make it easier to write and read formulas. An abbreviated wff is not a formal wff.

Example 10: $(((p_0) \rightarrow ((p_1) \wedge (p_2))) \rightarrow ((((p_{13}) \wedge (p_6)) \vee (p_{317})) \rightarrow (p_{26})))$

Analysis Abbreviating this wff, we can see it has length 5.

$$(p_0 \rightarrow p_1 \wedge p_2) \rightarrow ((p_{13} \wedge p_6) \vee p_{317} \rightarrow p_{26})$$

with lengths: 1, 1, 1, 1, 1, 1, 1; grouped as 2, 2; then 3, 3; then 4; then 5.

Exercises

1. Why do we introduce a formal language?

2. Identify which of the following are formal (unabbreviated) wffs:
   a. $(p_1) \vee \neg(p_2)$
   b. $((p_1) \rightarrow (p_2))$
   c. $((p_1 \vee p_2) \rightarrow p_2)$
   d. $(\neg(p_1)(p_2) \wedge (p_1))$
   e. $(\neg(\neg(p_1)) \vee \neg(p_1))$
   f. $((\neg(\neg(p_1))) \vee \neg(p_1))$
   g. $((\neg(\neg(p_1))) \vee (\neg(p_1)))$

3. Abbreviate the following wffs according to our conventions and then give the length of each.
   a. $(((\neg p_0) \wedge p_{13}) \rightarrow p_2)$
   b. $((((p_1) \rightarrow (p_2)) \wedge (\neg(p_2))) \rightarrow (\neg(p_1)))$
   c. $((((p_4) \wedge (p_2)) \vee (\neg(p_6))) \rightarrow ((p_7) \rightarrow (p_8)))$

4. Give an example (abbreviated is O.K.) of a wff that is:
   a. A conjunction, the conjuncts of which are disjunctions of either propositions or negated propositions.
   b. A conditional whose antecedent is a disjunction of negations and whose consequent is a conditional whose consequent is a conditional.

Aside: Defining a formal language without induction

Some logicians define the formal language of propositional logic as:

$L(\neg, \rightarrow, \wedge, \vee, p_0, p_1, \ldots)$ is the smallest collection containing (p_i) for each $i = 0, 1, 2, \ldots$ and closed under the formation of wffs, that is, if A, B are in the collection, so are $(\neg A)$, $(A \rightarrow B)$, $(A \wedge B)$, and $(A \vee B)$.

One collection is said to be smaller than another if it is contained in or equal to the other.

This definition seems to require us to accept that there are completed infinite collections. Yet when these logicians want to prove anything about the formal language, they still have to give a definition of the length of a wff, and that amounts to giving an inductive definition of the language in addition to this one.

C. Realizations and Semi-Formal Languages

The formal language gives us the forms of the propositions we will study. A wff such as $p_0 \wedge \neg p_1$ is neither true nor false; it is the skeleton of a proposition. Only when we fix on a particular interpretation of the formal connectives and then assign propositions to the variables, such as "p_0" stands for "Ralph is a dog" and "p_1" stands for "Four cats are sitting in a tree", do we have a semi-formal proposition, "Ralph is a dog $\wedge$ $\neg$ (four cats are sitting in a tree)" that can be viewed as having a truth-value. We may read this as "Ralph is a dog and it's not the case that four cats are sitting in a tree" so long as we remember that we've agreed that all "and" and "it's not the case that" mean will be captured by the interpretations we'll give for $\wedge$ and $\neg$.

Realizations and semi-formal languages A *realization* is an assignment of propositions to some or all of the propositional variables. The *realization of a formal wff* is the formula we get when we replace the propositional variables appearing in the formal wff with the propositions assigned to them; it is a *semi-formal wff*. The *semi-formal language* for the realization is the collection of realizations of the formal wffs.

Example 11: Here is a realization:

p_0	"Ralph is a dog"
p_1	"Four cats are sitting in a tree"
p_2	"Four is a lucky number"
p_3	"Dogs bark"
p_4	"Juney is barking loudly"
p_5	"Juney is barking"
p_6	"Dogs bark"
p_7	"Ralph is barking"
p_8	"Cats are nasty"
p_9	"Ralph barks"
p_{47}	"Howie is a cat"
p_{312}	"Bill is afraid of dogs"
p_{317}	"Bill is walking quickly"
p_{4318}	"If Ralph is barking, then he will catch a cat"
p_{4319}	"Ralph is barking"

Analysis We don't need to realize all the propositional variables. And we can assign the same proposition to more than one variable. In this example, the realization of p_7 is the same as the realization of p_{4319}. And the realization of: $((p_0) \to ((p_1) \wedge (p_2)))$ is:

((Ralph is a dog) $\to$ ((Four cats are sitting in a tree) $\wedge$ (Dogs bark)))

The propositions we assign to the propositional variables are meant to be the simplest ones we can begin with. We say they're *atomic* because we are not concerned with their internal structure, since they involve no logical symbols. So the assignment of "If Ralph is barking, then he will catch a cat" to p_{4318} is a bad choice, for it will not allow us to make an analysis based on the form of that proposition in relation to what we assign to p_7, "Ralph is barking".

Semi-formal wffs inherit the terminology of the formal wffs they realize: atomic, compound, conjunction, etc. We can use the same metavariables $A, B, C, A_0, A_1, A_2, \ldots$ for formal wffs, semi-formal wffs, or propositions in English, trusting to context to make clear which are meant.

Exercises
1. For Example 11, give the realization of each of the following (abbreviated) wffs.
    a. $((p_8 \wedge p_{4318}) \wedge p_7) \to p_1$
    b. $(p_0 \wedge p_1) \to p_2$
    c. $\neg(p_4 \wedge \neg p_5)$
    d. $p_3 \to \neg\neg p_6$
    e. $\neg(p_{312} \wedge p_7) \wedge \neg p_{317}$
    f. $p_{312} \wedge p_7 \to p_{317}$
2. The following wffs are (an abbreviations of) the realization of what formal wffs in the realization of Example 11?
    a. Ralph is a dog $\wedge$ $\neg$ (four cats are sitting in a tree)
    b. Bill is afraid of dogs $\wedge$ Ralph barks $\to$ Bill is walking quickly
    c. Juney is barking loudly $\to$ Juney is barking
    d. Ralph is a dog $\wedge$ dogs bark $\to$ Ralph barks
    e. Four cats are sitting in a tree $\wedge$ four is a lucky number $\to$
        $\neg$(Dogs bark $\to$ Howie is a cat)

Key Words	formal logic	compound wff
	propositional logic	unique readability of wffs
	$\wedge, \vee, \neg, \to$	realization
	formal language	semi-formal language
	atomic wff	

4 Classical Propositional Logic: Meaning

A.	Meaning	21
	Exercises	24
B.	Models	25
C.	Tautologies	26
	Exercises	28
D.	Semantic Consequence	29
	Exercises	32
E.	Semantically Equivalent Formulas	34
	Exercises	34

A. Meaning

We have established the formal syntax, which we use to establish semi-formal languages. The simplest wffs in a semi-formal language are sentences from English that we consider to be atomic propositions: they have no structure in terms of the formal connectives.

The compound wffs, however, aren't propositions until we agree on how we'll understand the formal connectives. When we were evaluating inferences in Chapter 2, the only semantic value we considered for the propositions that made up a compound was truth-value. We didn't talk about how likely it was that "Dick will wake up" is true, nor concern ourselves with the subject matter of "Suzy calls early". We considered only the truth-values of the parts, and then assumed, without ever saying it, that the truth-value of the compound depended on just those values. Let's make those assumptions explicit.

The Classical Abstraction The only semantic property of an atomic proposition we will take into account is its truth-value.

Compositionality The truth-value of a proposition is determined by its form and the semantic properties of its constituents.

These are simplifications, abstractions from the meanings and properties of atomic propositions and the connectives.

For wffs of the form $p \wedge q$, $p \vee q$, $p \rightarrow q$, and $\neg p$, where p and q are atomic, there is no form to p or q, so the truth-value of the whole must be a function of the just truth-values of the parts. But which functions? I'll let you convince yourself that the only reasonable choices for $\neg$ and $\wedge$ are given by the following tables, where I write T for "true" and F for "false".

p	¬p
T	F
F	T

If p is T, then ¬p is F;
if p is F, then ¬p is F.

p	q	p∧q
T	T	T
T	F	F
F	T	F
F	F	F

p ∧ q is T if both p and q are T; otherwise it's F.

For ∨ there are two choices, corresponding to *inclusive* "or" in English ("A or B or both") and *exclusive* "or" ("A or B but not both"). The choice is arbitrary. It's customary now to use the inclusive version:

p	q	p∨q
T	T	T
T	F	T
F	T	T
F	F	F

p ∨ q is T if either p or q is T; otherwise it's F.

The only real issue is what table to use for →. First I'll set it out and then explain why we choose this one.

p	q	p→q
T	T	T
T	F	F
F	T	T
F	F	T

The first two lines codify the direct way of reasoning with conditionals (p. 11). From "If Flo comes over to play, then Dick will wake up" and "Flo comes over to play", we can conclude "Dick will wake up". So if p → q is T and p is T, then q is T, which is the first line. And if the antecedent is true and the consequent false, we can't have the conditional true, for then we could use the direct way to conclude the consequent. So if p is T and q is F, the conditional has to be F, which is the second line.

The only real issue is what to do if the antecedent is F. Consider:

(1) If someone is convicted of murder, then he or she will go to prison.

This is true (in the United States). But if it's true, it must be true no matter who the person is. So "If Lee is convicted of murder, then he'll go to prison" is true, even though Lee has never been convicted of any crime and has never been in prison. So we need that if p is F and q is F, then p → q is T. But also "If Zeke is convicted of murder, he'll go to prison" is true. But Zeke is in prison because he was convicted of animal abuse, not murder. So we want that if p is F and q is T, then p → q is T. The evaluation of the conditional by this table allows us to classify conditionals such as (1) as true by ignoring cases where the antecedent is F. That is, p → q is T iff (p is F or q is T). So for (1) to be false, there'd have to be someone who was convicted of murder (p is T) and did not go to prison (q is F). We call the conditional T when the antecedent is F because "the antecedent doesn't apply".

Given any atomic propositions p and q, we have a semantic analysis of ¬p, p→q, p∧q, and p∨q. But what about:

(2) ¬(¬(Ralph is a dog))

Shall we say this is true just in case "¬(Ralph is a dog)" is false? Or should we follow what we do in ordinary speech and say that more than one negation is just for emphasis? When someone says, "I don't want no broccoli" we don't understand that as a request for some broccoli. If we were to count the number of negations in determining the truth-value, it would amount to giving some semantic value to the form. Mixing syntax and semantics that way might be interesting, but it will not be simple. So we'll assume that the form of "(¬(Ralph is a dog))" does not matter in determining whether (2) is true. We treat (2) as having the form ¬A and consider only the truth-value of A. Hence, the table for ¬ should apply, and (2) is T iff "(¬(Ralph is a dog))" is F. Similarly, we'll treat "(Ralph is a dog ∧ dogs bark) → Ralph barks" as having the form (A ∧ B) → C and concern ourselves only with whether A, B, and C are T or F.

The Division of Form and Content   If two propositions have the same semantic properties, then they are indistinguishable in any semantic analysis regardless of their form.

So the truth-value of the whole is a function of just the truth-values of its parts, and we can use the truth-tables we devised for atomic propositions for compound propositions.

The classical truth-tables

A	¬A	A	B	A∧B	A	B	A∨B	A	B	A→B
T	F	T	T	T	T	T	T	T	T	T
F	T	T	F	F	T	F	T	T	F	F
		F	T	F	F	T	T	F	T	T
		F	F	F	F	F	F	F	F	T

Note that to use these tables it is crucial that each wff can be read in one and only one way, which we proved in Chapter 3. Note also that I've used "and", "or", "not", and "if . . . then . . ." to explain these tables. This isn't circular. We're not defining or giving meaning to those words but to ∧, ∨, ¬, and →. I have to assume you understand the ordinary English words.

Suppose now we have a semi-formal language. We agree that the sentences assigned to the propositional variables are propositions. In addition, we have the truth-tables for assigning truth-values to compound formulas in terms of the truth-values of their parts. Are we then to assume that every well-formed formula of the semi-formal language is a proposition? Consider:

¬¬¬¬¬¬¬¬¬¬¬¬¬¬¬ (Ralph is a dog)

We might take this to be a formalization of a common language proposition, though it's unlikely we'd use it in our reasoning. But what if there are 613 negation signs in front of "(Ralph is a dog)"? It seems harmless to accept such a formula as a proposition, even though we might never use it. And to try to establish further criteria for what formulas of the semi-formal language are to be propositions beyond saying that they are well-formed or perhaps shorter in length than some specific limit involves us in a further semantic analysis not already taken into account by the assumptions we've adopted.

Form and Meaningfulness What is grammatical and meaningful is determined solely by form and what primitive parts of speech are taken as meaningful. In particular, for a semi-formal language every well-formed formula will be taken to be a proposition.

Exercises

1. Why do we take A→ B to be true if A is false?
2. Which of the following italicized phrases could reasonably be construed as a truth-functional connective in the given sentence?
   a. *If* sodium burns, *then* it is not a metal.
   b. You can't make an omelette *without* breaking eggs.
   c. *If* Suzy passed that test, *then* I'm a monkey's uncle.
   d. The sky is blue *because* the sun is yellow.
   e. Earl climbed the ladder *and then* painted the eaves.
   f. Ralph went to the movie *and* bought popcorn.
   g. *If* the moon is made of green cheese, *then* 2 + 2 = 4.
   h. *Either* 2 + 2 = 4 *or* 4 + 4 = 7.
   i. *If* Donald Trump were a dog, *then* he would bark too much.
   j. Dick *believes* that Spot bit Puff.
   k. There is *no* one who can lift 500 kilograms.
   l. *Neither* Ralph *nor* Dusty is a dog.
3. Assign truth-values to the atomic propositions in the realization of Example 11, p. 19 and evaluate the truth-value of the realizations of the following wffs:
   a. $(p_1) \vee \neg(p_2)$
   b. $\neg(p_4 \wedge \neg p_5)$

c. $p_3 \to \neg\neg p_6$
d. $\neg(p_{312} \land p_7) \land \neg p_{317}$
e. $p_{312} \land p_7 \to \neg p_{317}$
f. $((((p_1) \to (p_2)) \land (\neg(p_2))) \to (\neg(p_1)))$
g. $((((p_4) \land (p_2)) \lor (\neg(p_6))) \to ((p_7) \to (p_8)))$
h. $((p_8 \land p_{4318}) \land p_7) \to p_1$
i. $(p_0 \land p_1) \to p_2$

4. For each of the following (abbreviated) semi-formal wffs, if possible assign truth-values to the atomic propositions in it that will make the entire wff be F.
   a. Ralph is barking → cats are nasty
   b. (Ralph is a dog ∨ dogs bark) ∧ ¬(Juney is barking)
   c. Ralph is barking ∧ cats are nasty → Ralph is barking
   d. Dogs bark ∨ ¬(dogs bark)   [as a realization of $p_6 \lor \neg p_6$]
   e. Dogs bark ∨ ¬(dogs bark)   [as a realization of $p_6 \lor \neg p_3$]
   f. ¬(cats are nasty → ¬¬(cats are nasty))
   g. ((Ralph is a dog ∨ dogs barks) ∧ ¬(Ralph is dog)) → dogs bark

C. Models

To evaluate inferences in Chapter 2 we had to decide whether there was a way in which the premises could be true and conclusion false. A "way" is a way the world could be. Here, the only connection between the world and a semi-formal language is the truth-values of the atomic propositions. So a way the world could be—as far as we are concerned here—is a determination of which atomic propositions are true. We call that a *valuation*. Whether an atomic proposition is true or false is not for us as logicians to decide. What we are concerned with is the range of ways the world could be, not identifying which assignment is "correct".

Given an assignment of truth-values to the atomic propositions, the truth-values of all the propositions in the semi-formal language are determined by the truth-tables.

Valuations and models A *valuation* v is an assignment of truth-values to the atomic propositions of the semi-formal language. A *model* is the semi-formal language, a valuation, and the extension of that to all compound wffs by the truth-tables.

Here is a diagram that gives the idea of a model, where "real(p_i)" stands for the realization of the propositional variable p_i.

26 An Introduction to Formal Logic

$$L(\neg, \rightarrow, \wedge, \vee, p_0, p_1, \ldots)$$
$$\downarrow \quad \text{realization}$$
$$\{\text{real}(p_0), \text{real}(p_1), \ldots, \text{propositions formed from these using } \neg, \rightarrow, \wedge, \vee\}$$
$$\downarrow \quad \text{v plus truth-tables}$$
$$\{\mathsf{T}, \mathsf{F}\}$$

If v assigns T to A, we write v(A) = T or v ⊨ A and say that A is *true in the model* or v *validates* A. We write v(A) = F or v ⊭ A if A is false in the model. We can use $M, M_0, M_1, \ldots$ to stand for models.

D. Tautologies
We formalize "Ralph is a dog or Ralph is not a dog" as

Ralph is a dog ∨ ¬ Ralph is a dog

This is true in every model regardless of whether Ralph is a dog or not According to our restricted understanding of a way the world is, this means it's true in every way the world could be. Since we want to talk about all ways the wffs could be true or false, we need to assume the following.

Sufficiency of the collection of models For any realization, any assignment of truth-values to the atomic propositions defines a model.

Then we are justified in making the following definition.

Classical tautologies   A wff is a *formal tautology* or *valid* iff in every model its realization is evaluated as true.
   A semi-formal proposition is a tautology iff it is the realization of a wff that is a tautology.
   A proposition in ordinary English is a tautology if there is a good formalization of it that is a tautology.
   A scheme of wffs is a tautology iff every wff of that form is a tautology.
   We write ⊨A to mean that A is a tautology.
   A tautology is sometimes called a *valid* wff or proposition.

Example 1: *If Ralph is a dog or Howie is a cat, and Ralph is not a dog, then Howie is a cat.*
Analysis Is this a tautology? An obvious formalization of it is:

((Ralph is a dog ∨ Howie is a cat) ∧ ¬(Ralph is not a dog)) → Howie is a cat

This has the form $((p_1 \vee p_2) \wedge \neg p_1) \rightarrow p_2$. Is every realization of that in any model going to be true? We can check by surveying all ways we could assign truth-values to its parts by making a table.

p_1	p_2	$p_1 \vee p_2$	$\neg p_1$	$(p_1 \vee p_2) \wedge \neg p_1$	$((p_1 \vee p_2) \wedge \neg p_1) \rightarrow p_2$
T	T	T	F	F	T
T	F	T	F	F	T
F	T	T	T	T	T
F	F	F	T	F	T

In this we don't mean that p_1 is true: p_1 is just a symbol. We mean that given any realizations of p_1 and p_2 that have these truth-values, we get this table. And no matter what those assignments are, the entire wff is evaluated as true. So we can say that the example is true due to its form—at least relative to the assumptions we've made about how to interpret "or", "and", "not" and "if . . . then . . .".

Example 2: *If not both Ralph is a dog and Ralph barks, then if Ralph barks, he is a dog.*
Analysis Is this a tautology? An obvious formalization of it is:

$\neg$ (Ralph is a dog $\wedge$ Ralph barks) $\rightarrow$ (Ralph barks $\rightarrow$ Ralph is a dog)

This has the form $\neg (p_1 \wedge p_2) \rightarrow (p_2 \rightarrow p_1)$. We can make up a table.

p_1	p_2	$p_1 \wedge p_2$	$\neg(p_1 \wedge p_2)$	$p_2 \rightarrow p_1$	$\neg(p_1 \wedge p_2) \rightarrow (p_2 \rightarrow p_1)$
T	T	T	F	T	T
T	F	F	T	T	T
F	T	F	T	F	F
F	F	F	T	T	T

We see that if p_1 is F and p_2 is T, the whole wff comes out F. So the formal wff is not a tautology, and hence the example isn't a tautology.

Making up the table for Example 2 takes a while. For a wff with that has six propositional variables, it would take a lot longer. And for a wff that has 47 variables in it, you'd run out of memory trying to make a table for it on a computer. For a wff with n propositional variables in it, the table will have 2^n rows. But there is a shortcut.

Example 3: $\neg (A \wedge B) \rightarrow (B \rightarrow A)$
Analysis This is the form of the last example. We know that it can be F iff the antecedent, $\neg (A \wedge B)$ is T and the consequent $(B \rightarrow A)$ is F. So we try to come up with a valuation that would do that:

```
           ¬ ( A ∧ B ) → ( B → A )   is F
    iff         T            F
    iff        A ∧ B       B   A
                 F         T   F
```

This is a falsifying assignment. So this is not a scheme of tautologies.

Example 4: $((A \wedge B) \to C) \to (A \to (B \to C))$

Analysis Is this a scheme of tautologies? We try to falsify it.

$((A \wedge B) \to C) \to (A \to (B \to C))$ is F

iff   T   F

iff   A $B \to C$
                           T   F

iff   B   C
                           T   F

But if A is T, B is T, and C is F, then $(A \wedge B) \to C$ is F. So there is no way to falsify the scheme. This is a scheme of tautologies.

Example 5: Ralph is a dog $\wedge \neg$ (Ralph is a dog)

Analysis This is false in any model. A proposition or formal wff that is evaluated as false regardless of the truth-values of its atomic constituents is an *anti-tautology*.

Exercises

1. What is a model?
2. a. What is a tautology in the formal language?
   b. What is a tautology of the semi-formal language?
   c. What is a tautology in ordinary English?
3. a. Exhibit two formal tautologies.
   b. Exhibit two semi-formal tautologies.
   c. Exhibit two informal tautologies.
   d. Exhibit two formal anti-tautologies.
   e. Exhibit two informal anti-tautologies.
4. Determine whether the following (abbreviated wffs) are tautologies (p, q, and r are used in place of p_1, p_2, and p_3).
   a. $(p \wedge q) \to (p \vee q)$
   b. $(p \vee q) \to (p \wedge q)$
   c. $(p \to q) \to (\neg p \vee q)$
   d. $(p \wedge q) \to p$
   e. $[\neg (p \wedge q) \wedge p] \to \neg q$
   f. $\neg q \to [\neg (p \wedge q) \wedge p]$
   g. $[(p \to q) \wedge (\neg p \to r)] \to (q \vee r)$
   h. $\neg (p \wedge \neg \neg p)$
   i. $\neg \neg p \vee \neg \neg \neg p$
   j. $((p \vee q) \wedge \neg p) \to q$
   k. $(q \vee \neg q) \to q$
   l. $q \to (q \vee \neg q)$
   m. $q \to \neg q$
   n. $\neg \neg q \to q$

o. ¬(¬p ∨ ¬¬p) ∨ p
p. p → (p ∨ q)
q. ¬p → (p → q)
r. p → (q → p)
s. ((p ↔ ¬q) ↔ ¬p) ↔ q
t. [q ↔ (r → ¬p)] ∨ [(¬q → p) ↔ r]
u. ¬(p ∧ ¬p)   *law of noncontradiction*
v. p ∨ ¬p   *excluded middle*, *tertium non datur* (a third way is not given)

E. Semantic Consequence

An inference is valid iff there is no way for its premises to be true and conclusion false at the same time. So according to our restricted understanding of a way the world can be, an inference is valid means there is no model in which the premises are true and conclusion false. So we make the following definition, where the Greek capital letters Γ (gamma), Δ (delta), and Σ (sigma) stand for collections of wffs, semi-formal propositions, or propositions, depending on the context.

Semantic consequence   For a collection of formal wffs Γ and a wff A, Γ *therefore* A is *valid* means there is no model in which the realizations of all the wffs in Γ are true and the realization of A is false. In that case we say that A is a *semantic consequence* of Γ, or that the pair Γ, A is a *semantic consequence*.

A semi-formal inference is valid iff it is the realization of a formal inference that is valid. An inference in ordinary English is valid if there is a good formalization of it which is valid.

If Γ *therefore* A is valid, we write Γ ⊨ A; it it is not valid, we write Γ ⊭ A.

Example 6:   *Either Ralph is a dog or Howie is a duck. No way is Howie a duck. Therefore, Ralph is a dog.*
Analysis   If we formalize "either . . . or . . ." as inclusive classical disjunction and formalize "no way" as classical negation, we get the semi-formal inference:

   Ralph is a dog ∨ Howie is a duck
   ¬ (Howie is a duck)
   Therefore, Ralph is a dog

This has the form of the formal inference:\

   $p_7 \vee p_{10}$ (We can use a line to separate the premises from
   $\underline{\neg p_{10}}$ the conclusion rather than writing "therefore".)
   p_7

Again, we can use a truth-table to decide.

30 An Introduction to Formal Logic

p_7	p_{10}	$p_7 \vee p_{10}$	$\neg p_{10}$
T	T	T	F
T	F	T	T
F	T	T	F
F	F	F	T

But here we don't look at the last column to determine whether the inference is valid. We look at rows in which the premises are T, which is just the second one. Then we look to see if the conclusion is T in that row, which it is. So the inference is valid. This confirms that Excluding Possibilities (p. 7) is valid.

Example 7: $p \to q$
              $\underline{\neg p \to r}$
              $q \vee r$

Analysis I've written p for p_1, q for p_2, and r for p_3 to make this easier to read.

p	q	r	$\neg p$	$p \to q$	$\neg p \to r$	$q \vee r$
T	T	T	F	*T*	*T*	T
T	T	F	F	*T*	*T*	T
T	F	T	F	F	T	T
T	F	F	F	F	T	F
F	T	T	T	*T*	*T*	T
F	T	F	T	T	F	T
F	F	T	T	*T*	*T*	T
F	F	F	T	T	F	F

I've put in italics and underlined where both premises are T. In each of those rows the conclusion is T. So the formal inference is valid: there's no way for the premises to be T and conclusion F.

Example 8: $\underline{p \vee q}$
              $p \wedge q$

Analysis

p	q	$p \vee q$	$p \wedge q$
T	T	T	T
T	F	T	F
F	T	T	F
F	F	F	F

In the second row the premise is T but the conclusion is F. So the formal inference is not valid.

The condition for $A \vDash B$ is that in every model in which A is T, so is B. But that means that in every model, $\vDash A \rightarrow B$. The condition for $\vDash A \rightarrow B$ is that $A \rightarrow B$ is true in every model, so in any model in which A is T, so is B. In that case $\vDash A \rightarrow B$. We've just shown the formal version of Reasoning from Hypotheses (p. 11).

The Semantic Deduction Theorem
$\vDash A \rightarrow B$ iff $A \vDash B$

Example 9: $[\neg(p \wedge q) \wedge p] \rightarrow \neg q$
Analysis This is a tautology (Exercise 4.e above). So by the Semantic Deduction Theorem, the following is a valid inference:

$$\frac{\neg(p \wedge q) \wedge p}{\neg q}$$

Because of our assumption of the division of form and content, any inference that involves only propositional variables that is valid (or invalid) is also valid (or invalid) for compound wffs. Thus, the following is a scheme of valid inferences:

$$\frac{\neg(A \wedge B) \wedge A}{\neg B}$$

When there are several premises for an inference, we can list them as separated by commas rather than give each a separate line. So, for example, the following is a scheme of valid formal inferences:

$$\frac{(A \rightarrow B), (\neg A \rightarrow C)}{(B \vee C)}$$

The following theorem summarizes some basic properties of the semantic consequence relation, using notation from set theory (Appendix 2).

Theorem 2 Basic properties of the semantic consequence relation

a. $A \vDash A$.

b. If $\vDash A$, then $\Gamma \vDash A$.

c. If $A \in \Gamma$, then $\Gamma \vDash A$.

d. If $\Gamma \vDash A$ and $\Gamma \subseteq \Delta$, then $\Delta \vDash A$.

e. *Transitivity* If $\Gamma \vDash A$ and $A \vDash B$, then $\Gamma \vDash B$.

f. If $\Gamma \vDash A$ and $\Delta \cup \{A\} \vDash B$, then $\Gamma \cup \Delta \vDash B$.

g. If $\Gamma \cup \{A_1, \ldots, A_n\} \vDash B$ and $\Gamma \vDash A_1, \ldots, \Gamma \vDash A_n$, then $\Gamma \vDash B$.

h. *The Semantic Deduction Theorem* $A \vDash B$ iff $\vDash A \rightarrow B$.

Review and Summary
We began by setting up a formal language to make precise the forms of the propositions we'll study. We based that on abstracting the four connectives

32   An Introduction to Formal Logic

"it's not the case that", "if . . . then . . .", "and", "or". A semi-formal language is what we use for reasoning, which is English with the formal propositional connectives ¬, →, ∧, ∨. Then we made the assumption that the only aspects of a proposition that are of concern to logic are its truth-value and its form as built up from sentence connectives, and that only these contribute to the truth-value of the whole. This led to the truth-tables for the formal connectives. Then we explained what it means for a proposition to be true due to its form by reference to the formal language, and we saw what it means for one proposition to follow from one or more others due solely to their form relative to the assumptions we've made about form and meaning. Now we have a logic.

Classical propositional logic   *Classical propositional logic* comprises the formal language along with the definitions of realization, models, tautology, and semantic consequence.

Classical propositional logic is often called the *classical propositional calculus*, abbreviated as **PC**.

Exercises
1. a. What is a semantic consequence in classical propositional logic?
    b. What is the relation of semantic consequence to "follows from"?
    c. Give a formal semantic consequence.
    d. Give a semi-formal semantic consequence.
    e. Give an informal semantic consequence.
2. Formalize the following from Chapter 2 and show that they are formally valid.
    a. Excluding possibilities (*disjunctive syllogism*)
    b. The Direct Way of Reasoning with Conditionals (*modus ponens*)
    c. The Indirect Way of Reasoning with Conditionals (*modus tollens*)
    d. Reasoning in a Chain with Conditionals (*transitivity of the conditional*)
3. Formalize the following from Chapter 2 and show that they are formally invalid.
    a. Affirming the Consequent
    b. Denying the Antecedent
4. Show that Reducing to the Absurd (*reductio ad absurdum*) is valid.
    $$\frac{A \to B,\ A \to \neg B}{\neg A}$$
5. Evaluate whether the following are schemes of valid inferences.
    a. $\dfrac{A \wedge B}{A \vee B}$
    b. $\dfrac{A \wedge \neg A}{B}$ ("everything follows from a contradiction")
    c. $\dfrac{(A \wedge \neg\neg A) \vee B}{\neg B}$

Chapter 4 Classical Propositional Logic: Meaning 33

d. $\dfrac{A \to \neg B,\ B \wedge \neg C}{A \to C}$

e. $\dfrac{A \to \neg\neg B,\ \neg C \vee A,\ C}{B}$

6. Here are some tautologies that people talk about when they use logic. Convert each to a scheme of valid inferences using the Semantic Deduction Theorem.

  a. $\left.\begin{array}{l}\neg\neg A \to A \\ A \to \neg\neg A\end{array}\right\}$ laws of double negation

  b. $\left.\begin{array}{l}(A \wedge B) \to A \\ (A \wedge B) \to B\end{array}\right\}$ laws of simplification for conjunction

  c. $(A \wedge A) \leftrightarrow A$ principle of tautology for conjunction

  d. $(A \wedge B) \to (B \wedge A)$ commutativity of conjunction

  e. $((A \wedge B) \wedge C) \leftrightarrow (A \wedge (B \wedge C))$ associativity of conjunction

  f. $\left.\begin{array}{l}A \to (A \vee B) \\ B \to (A \vee B)\end{array}\right\}$ laws of addition for disjunction

  g. $(A \vee A) \leftrightarrow A$ principle of tautology for disjunction

  h. $(A \vee B) \to (B \vee A)$ commutativity of disjunction

  i. $((A \vee B) \vee C) \leftrightarrow (A \vee (B \vee C))$ associativity of disjunction

  j. $\left.\begin{array}{l}A \vee (B \wedge C) \leftrightarrow ((A \vee B) \wedge (A \vee C)) \\ A \wedge (B \vee C) \leftrightarrow ((A \wedge B) \vee (A \wedge C))\end{array}\right\}$ distribution laws

  k. $\left.\begin{array}{l}\neg(A \wedge B) \leftrightarrow (\neg A \vee \neg B) \\ \neg(A \vee B) \leftrightarrow (\neg A \wedge \neg B)\end{array}\right\}$ De Morgan's laws

  l. $A \to A$ identity

  m. $(\neg A \to A) \to A$ Clavius' law, consequentia mirabilis

  n. $\left.\begin{array}{l}A \to (B \to A) \\ \neg A \to (A \to B)\end{array}\right\}$ the paradoxes of material implication

  o. $(A \to B) \to ((\neg A \to B) \to B)$

  p. $\left.\begin{array}{l}A \to (B \to B) \\ (A \wedge \neg A) \to B\end{array}\right\}$ the paradoxes of strict implication

  q. $\left.\begin{array}{l}((A \to B) \wedge (B \to C)) \to (A \to C) \\ (A \to B) \to ((B \to C) \to (A \to C))\end{array}\right\}$ transitivity of $\to$

  r. $((A \wedge B) \to C) \to (A \to (B \to C))$ exportation

  s. $(A \to (B \to C)) \to ((A \wedge B) \to C)$ importation

  t. $(A \to (B \to C)) \to (B \to (A \to C))$ interchange of premises

  u. $(A \to (B \wedge C)) \to ((A \to B) \wedge (A \to C))$

v. $(A \rightarrow (B \vee C)) \rightarrow ((A \rightarrow B) \vee (A \rightarrow C))$

w. $(A \rightarrow (B \rightarrow C)) \rightarrow ((A \rightarrow B) \rightarrow (A \rightarrow C))$

7. Prove Theorem 2 on the basic properties of the semantic consequence relation.

E. Semantically Equivalent Formulas

When is $A \rightarrow B$ true? Just in case A isn't T and B is F. That's the same condition for $\neg (A \wedge \neg B)$ to be true. So in every model the realizations of those two wffs will both be true or both be false.

Semantically equivalent formulas Two wffs are *semantically equivalent* iff in every model they have the same semantic properties.

For classical propositional logic that means that given any model, either both are true or both are false.

Theorem 3 Each of these pairs are semantically equivalent.

$A \rightarrow B$ and $\neg (A \wedge \neg B)$

$A \rightarrow B$ and $\neg A \vee B$

$A \wedge B$ and $\neg (\neg A \vee \neg B)$

$A \wedge B$ and $\neg (A \rightarrow \neg B)$

$A \vee B$ and $\neg (\neg A \wedge \neg B)$

$A \vee B$ and $\neg A \rightarrow B$

You can prove this by using the truth-tables for the two (schemes of) wffs.

We based classical propositional logic on the four formal connectives: $\neg, \rightarrow, \wedge, \vee$. But we could have started with just $\neg$ and $\wedge$, and then set:

$A \rightarrow B \equiv_{Def} \neg (A \wedge \neg B)$

$A \vee B \equiv_{Def} \neg (\neg A \wedge \neg B)$

Perhaps more economical, but not necessarily clearer.

Exercises

1. Show that any wffs A and B are semantically equivalent iff $\vDash A \leftrightarrow B$.
2. Show that starting with just $\neg$ and $\vee$ we can define $\rightarrow$ and $\wedge$.
3. Show that starting with just $\wedge, \vee$, and $\rightarrow$ we can't define $\neg$.
4. Define the connective *nand* (not-and, also called the "Sheffer stroke") :

   $A \mid B$ is T iff not both A and B are T

   Show that $\neg A$ is semantically equivalent to $A \mid A$, and $A \rightarrow B$ to $A \mid \neg B$.

5. Define a connective to formalize "neither . . . nor . . .":

   $A \downarrow B$ is true iff both A and B are false

   Show that $\neg A$ is semantically equivalent to $A \downarrow A$, and $A \wedge B$ is equivalent to $\neg A \downarrow \neg B$.

Key Words

- Classical Abstraction
- Compositionality
- classical truth-tables
- inclusive "or"
- exclusive "or"
- The Division of Form and Content
- Form and Meaningfulness
- valuation
- model
- sufficiency of collection of models
- tautology
- anti-tautology
- semantic consequence
- The Semantic Deduction Theorem
- classical propositional logic
- semantically equivalent formulas

5 Using Classical Propositional Logic

To use classical predicate logic to evaluate our reasoning, we first have to formalize the reasoning. But what counts as a good formalization? Here we'll look at some examples in order to come up with criteria.

Example 1: *George is a duck and Ralph is a dog. Therefore Ralph is a dog.*
Analysis The informal inference is valid. If we formalize the premise as "George is a duck ∧ Ralph is a dog", taking "George is a duck" and "Ralph is a dog" as atomic, the semi-formal inference will be valid, too. Were we to take the premise as atomic, the formalization would be "p therefore q", which is not valid. That's wrong.
   An informally valid inference should be formalized as a valid one.

Example 2: *All dogs bark. Ralph is a dog. So Ralph barks.*
Analysis This is valid. And so is "All swans are white. Cecil is a swan. So Cecil is white." You can make up more examples to see that this example as valid due to its form. But there is no part of any of the propositions in it that we can formalize as a propositional connective. So its form in classical propositional logic is "p, q, therefore r", which is not valid. Hence, there is no good formalization of the example.

Example 3: *Ralph is a dog puppet. So Ralph is a dog.*
Analysis This isn't valid: the premise is true and conclusion is false. So we can't formalize the premise as "Ralph is a dog ∧ Ralph is a puppet" because from that the conclusion does follow. There's no good formalization of this.
   An informally invalid inference should not be formalized as a valid one.

Example 4: *Ralph is a dog or Ralph is not a dog.*
Analysis Informally this is true due to its form, which seems to depend only on the words "and" and "not". So a good formalization should show that. And "Ralph is a dog ∨ ⇁ (Ralph is a dog)" does, for that is a formal tautology.
   An informal tautology should be formalized as a tautology.

Example 5: *Ralph is a dog and Ralph is not a dog.*
Analysis I can imagine someone saying this after seeing the picture of Ralph on the back cover of this book. And she might think it's true. But if so, she's using "and" and "not" differently than we interpret them in classical propositional logic, for we would formalize this as "Ralph is a dog ∧ ⇁ (Ralph is not a dog)" which is an anti-tautology.
   An informal anti-tautology should be formalized as an anti-tautology.

Example 6: *Ralph is a dog but Howie is a cat.*
Analysis The connective "but" isn't one of the four connectives we've chosen to formalize. So should we take the example as atomic? No, because both "Ralph is a dog" and "Howie is a cat" follow from it, and we couldn't respect the validity of that

Chapter 5 Using Classical Propositional Logic 37

inference. The word "but" is used as "and" plus some sense of surprise. But that idea of surprise doesn't affect our reasoning, or at least we can argue that it needn't in the context of the assumptions of classical propositional logic. So we can formalize the example as "Ralph is a dog ∧ Howie is a cat".

Example 7 Spot is a dog but Puff is not a dog.
Analysis In accord with what we did in the last example, we should formalize this as "Spot is a dog ∧ ¬ (Puff is a dog)". We don't want to formalize haphazardly, inventing anew each time we encounter a proposition.
   Our formalizations should be regular, following some conventions.

Example 8: If Ralph is a dog but George is a duck, then George is a duck.
Analysis In parity with what we did in the last two examples, we can formalize this as:

   (Ralph is a dog ∧ George is a duck) → George is a duck

This is a tautology, since it can be a realization of the formal tautology:

(a) $(p_1 \land p_2) \to p_2$

It could also be taken as a realization of $(p_{32} \land p_{47}) \to p_{32}$. But that doesn't matter, since that's semantically equivalent to (a).
   Sometimes people say that (a) is the "logical form" of the example, as if it were clear and simple to formalize the example as we did. Yet the following makes us doubt the aptness of always formalizing "but" as ∧ :

   If Donald Trump is honorable but Donald Trump lied, then Donald Trump is honorable.

Example 9: Ralph is a dog or Dusty is a horse and Howie is a cat.
        Therefore, Howie is a cat.
Analysis Without a context there is no preference for reading the premise as (A ∨ B) ∧ C or as A ∨ (B ∧ C). But the inference does provide context, for from the premise it's meant that "Howie is a cat" follows. Hence, we formalize the example:

   (Ralph is a dog ∨ Dusty is a horse) ∧ Howie is a cat
   Therefore, Howie is a cat

Example 10: Ralph is a dog and George is a duck and Howie is a cat.
Analysis We could formalize the example as either:

or
   (Ralph is a dog ∧ George is a duck) ∧ Howie is a cat
   Ralph is a dog ∧ (George is a duck ∧ Howie is a cat)

These are equivalent Our formal language imposes a precision that is neither observed nor needed in ordinary speech but does not contradict our intuition that the placement of parentheses (or commas) in the example is immaterial for our reasoning. In order to have a general rule for how to formalize a sequence of conjuncts, we can make a convention that a series of conjuncts will be associated to the left if there's no indication otherwise. That is, " A and B and C" will be formalized as (A ∧ B) ∧ C. Let's agree that:

   A ∧ B ∧ C abbreviates (A ∧ B) ∧ C

Example 11: *Spot is a dog and Puff is a cat.*
Analysis The formalization is straightforward:

(a)   Spot is a dog ∧ Puff is a cat

Equivalent to that is "Puff is a cat ∧ Ralph is a dog". Yet we choose (*) in order to *respect the grammar of the original, which includes the order of the parts*.

Example 12: *Ralph is a dog if he's not a puppet.*
Analysis We replace the indexical "he" with "Ralph". Then "if" introduces the antecedent of the conditional, so we can rewrite the example as:

   If Ralph is not a puppet, then Ralph is a dog

which we can formalize as:

   ¬ (Ralph is a puppet) → Ralph is a dog

Respecting the grammar of the original and the order of the parts can be only a guide, not a fixed rule.

Example 13: *Ralph is a dog only if he is not a puppet.*
Analysis In accord with how we understand "only if" (Chapter 2, p. 9), we formalize the example as: "Ralph is a dog → ¬ (Ralph is a puppet)".

Example 14: *Ralph is a dog if and only if he is not a puppet.*
Analysis In accord with how we understand "if and only if" in Chapter 2 (p. 9), we formalize the example as:

(Ralph is a dog → ¬ (Ralph is a puppet)) ∧ (¬ (Ralph is a puppet) → Ralph is a dog)

We'll make an abbreviation.

   $A \leftrightarrow B \equiv_{Def} (A \rightarrow B) \wedge (B \rightarrow A)$

We call "↔" the *biconditional*.

Example 15: *If Ralph is a dog, then he's a greyhound.*
      So if Ralph is not a greyhound, he's not a dog.
Analysis We can formalize the example as:

   Ralph is a dog → Ralph is a greyhound
   Therefore, ¬ (Ralph is a greyhound) → ¬ (Ralph is a dog)

This has form:

   $(p_1 \rightarrow p_2)$
   Therefore, $(\neg p_2 \rightarrow \neg p_1)$

Here classical propositional logic helps us see that the informal inference is valid.
   You can check that (¬B → ¬A) is semantically equivalent to (A → B). Deriving the one from the other is called *contraposition*. We call ¬B → ¬A the *contrapositive* of A → B.

Example 16: *If Ralph is a dog, then Ralph barks. Ralph barks.*
      So Ralph is a dog.

Analysis This is not valid (it's an example of Affirming the Consequent). If you had doubts about that, formalize the example as:

Ralph is a dog → Ralph barks
Ralph barks
Therefore, Ralph is a dog

This has the invalid form: p→ q , q therefore p .

Example 17: *Ralph is a puppet; he is not a cat.*
Analysis We need to replace the indexical "he" with "Ralph". Though there is no word that could be construed as equivalent to "and" in this example, semicolons, commas, and other punctuation are often used in English in place of connectives and should be formalized accordingly. We formalize the example as:

Ralph is a puppet ∧ ¬ (Ralph is a cat)

Example 18: *Ralph is a dog or he's a puppet.*
Analysis We've agreed to formalize "or" as inclusive disjunction, so we get:

Ralph is a dog ∨ Ralph is a puppet

Here that seems odd, for we suppose the alternatives are mutually exclusive. If that's what's intended (which we'd have to figure out from context), we should formalize:

(Ralph is a dog ∨ Ralph is a puppet) ∧ ¬ (Ralph is a dog ∧ Ralph is a puppet)

We define *exclusive disjunction*: (A ∨ B) ∧ ¬ (A ∧ B) .

Example 19: *Bring me an ice cream cone and I'll be happy.*
Analysis I understand this to mean "If you bring me an ice cream cone, then I'll be happy". How do I know? Because I speak English. So it would be wrong to formalize "and" here as ∧ . We have to be careful in formalizing even the four English connectives we started with.

Example 20: *Three faces of a die are even numbered.*
   Three faces of a die are not even numbered.
   So cats bark.
Analysis The two premises appear to contradict each other: p and ¬ p . If so, the inference is valid, since in no model could the premises both be true. But the example is not valid: both of the premises are true and the conclusion is false. The word "not" in the second premise does not negate "Three faces of a die are even numbered" but applies only to "even numbered". We can't formalize that kind of negation in classical propositional logic.

Example 21: *Ralph is not a dog because he's a puppet.*
Analysis It looks like "because" is used as a connective. But "because" introduces a reason or reasons for believing. The example is a different way to say "Ralph is a puppet *therefore* Ralph is not a dog". That's an inference, not a single proposition. We don't treat "because" and "therefore" and variants of those like "since" and "hence" as connectives because that would confuse inferences with propositions. So we formalize the example as:

Ralph is a puppet
Therefore, ⌐ (Ralph is a dog)

Example 22: *Wanda wonders whether Ralph is a puppet. And Ralph is a puppet.*
Analysis To formalize this we'd have to take "Ralph is a puppet" as atomic. But we'd also have to take "Wanda wonders whether Ralph is a puppet" as atomic since the truth-value of that does not depend on whether "Ralph is a puppet" is true. So there's no way to formalize that using a truth-functional connective for to do so we would not be able to recognize that we have an atomic proposition within another proposition.

Example 23: *Suzy took off her clothes and went to bed.*
Analysis The example was true last night, but "Suzy went to bed and took off her clothes" was false. Yet "Suzy took off her clothes ∧ Suzy went to bed" is equivalent to "Suzy went to bed∧ Suzy took off her clothes ". The example is using "and" as "and then", which we can't formalize in classical propositional logic because we don't take account of time. Nor can we take the example as atomic, since "Suzy went to bed" follows from it. So we can't formalize this example.

Example 24: *Every counting number is even or odd.*
Analysis The example is atomic. We cannot formalize "or" here by ∨ , because the example is not equivalent to "Every counting number is even or every counting number is odd".

Example 25: *If the moon is made of green cheese, then 2 + 2 = 4.*
Analysis The antecedent is false, the consequent is true, so according to the classical table for the conditional, the example is true. That seems odd if not downright wrong because it seems there should be some connection of meaning or subject matter between antecedent and consequent for a conditional to be true. To take account of such a semantic value in addition to truth-values, we'd need to develop a different logic—but that's another story, which you can read in my *Propositional Logics*.

Some call (A ∧ ⌐A) → B and A → (B → B) *paradoxes of classical propositional* logic. But there's nothing paradoxical about these being tautologies. We never intended to capture with the classical truth-table all there is to how we use "if . . . then . . ." . The table for → is just the best we have once we agree to consider only truth-values and form in evaluating compound propositions.

In summary, we have adopted the following criteria for good formalizations.

Criteria of Formalization
- The formalization respects the assumptions that govern our choice of primitive connectives and definition of truth in a model.
- A proposition is informally false due to its form iff its formalization is an anti-tautology.
- One proposition follows informally from another proposition or collection of other propositions iff its formalization is a formal semantic consequence of the formalizations of the other(s).

Chapter 5 Using Classical Propositional Logic 41

And we've adopted the following conventions for formalizing as guidelines, though not rules, for we saw in some examples context can make it inappropriate.

- A regular translation of certain words as connectives should be observed. More generally, formalizations should be regular in the sense that each proceeds in analogy with agreed-upon formalizations of other examples. This is the requirement of *parity of form*.
- A formalization should respect the grammar of the original, which includes the order of the parts.

Exercises
1. Show how we can formalize the following connectives:
   a. not both A and B
   b. A unless B
   c. B just in case A
2. Formalize each of the following or explain why it cannot be formalized in classical propositional logic.
   a. 7 is not even.
   b. 7 is not even or odd.
   c. If 7 is even, then 7 is not odd.
   d. Every number is even or odd.
   e. If logic is hard, then art history isn't hard.
   f. Either Tom and Dick will walk to the party on Saturday, or Suzy will drive.
   g If the play is sold out then Zoe will stay home and Dick will meet us tonight.
   h. Unless Dick gives Manuel a ride, Manuel won't go to the dance.
   i. Spot barked before Dick yelled.
   j. It's impossible that $2 + 2 \neq 4$.
   k. Anubis, you know, eats, sleeps, and barks at night.
   l. You can't make an omelette without breaking eggs.
   m. If acid and water are mixed and you do not wish to be burned, then you should be careful.
   n. If Ralph is a puppet, and Ralph barks, then Ralph is not a puppet.
3. Formalize each of the following inferences and evaluate whether it is valid.
   a. If cat owners' homes have fleas, then cats are nasty. If cat owners' homes smell bad, then cats are nasty. But cat owners' homes always have fleas or smell bad. So cats are nasty.
   b. If strawberries are red, then some color-blind people cannot see strawberries among their leaves. Strawberries are red. So some color-blind people cannot see strawberries among their leaves.
   c. The students are happy if and only if no test is given. If the students are happy, the professor feels good. But if the professor feels good, he won't feel like lecturing, and if he doesn't feel like lecturing, a test is given. Therefore, the students are not happy.

d. Dogs are a man's best friend. A friend is loyal. Hence, dogs are loyal.
e. Tom is Polish, and it's not the case that Tom is from New York or Virginia. If Tom is from Syracuse, then Tom is from New York or Virginia. Therefore, Tom is not from Syracuse.
f. If Ralph is a cat, then Ralph meows. Ralph is not a cat. Therefore, Ralph does not meow.
g. If you know some logic, you are either very bright or you study very hard. You study very hard. You are very bright. Therefore, you know some logic.
h. The government is going to spend less on health and welfare.
If the government is going to spend less on health and welfare, then either the government is going to cut the Medicare budget or the government is going to slash spending on housing.
If the government is going to cut the medicare budget, the elderly will protest.
If the government is going to slash spending on housing, then advocates for the poor will protest.
Therefore, the elderly will protest or advocates for the poor will protest.
i. Ralph is a dog.
Therefore, Ralph is a dog.
j. Either the moon is made of green cheese or 2 + 2 = 4.
The moon is not made of green cheese.
Therefore, 2 + 2 = 4.
k. Either the moon is made of green cheese or 2 + 2 = 5.
The moon is not made of green cheese.
Therefore, 2 + 2 = 5.
l. All men are mortal. Socrates is a man.
Therefore, Socrates is mortal.

Key Words formalizing exclusive disjunction
exclusive disjunction biconditional
criteria of formalization only if
parity of form contrapositive
associating to the left

6 Proving

A. Proving 43
Exercises 46
B. Deriving Consequences from Hypotheses 46
Exercises 48

A. Proving

To determine whether a (semi-formal) wff is a tautology we can use truth-tables. This is the semantic approach to logic, based on how we understand truth in a model.

Alternatively, we can start with some basic wffs we feel everyone will agree are tautologies, which we all *axioms*. Then we give a method for deriving more wffs from those via rules. This is the syntactic approach to formal logic. Unlike the semantic approach, where there is basically just one way to develop classical propositional logic (via truth-tables), there are many ways to develop propositional logic syntactically, depending on which wffs we take as basic and what rule(s) we adopt to derive more.

To have a formal proof system, we have to specify the axioms and the rule(s). It's customary and easier to give the axioms as schemes of wffs: each wff that is an instance of the scheme is an axiom.

An axiom system for classical propositional logic

1. $\neg A \to (A \to B)$
2. $B \to (A \to B)$
3. $(A \to B) \to ((\neg A \to B) \to B)$
4. $(A \to (B \to C)) \to ((A \to B) \to (A \to C))$
5. $A \to (B \to (A \wedge B))$
6. $(A \wedge B) \to A$
7. $(A \wedge B) \to B$
8. $A \to (A \vee B)$
9. $B \to (A \vee B)$
10. $(A \to C) \to ((B \to C) \to ((A \vee B) \to C))$

rule $\dfrac{A,\ A \to B}{B}$ *modus ponens*

---

44   An Introduction to Formal Logic

Proof A *formal proof* or *derivation of* A is a sequence $A_1, \ldots, A_n$, where A_n is A and each A_i is either an axiom or is a derived from one or more of the preceding A_j's by one of the rules. In that case, we say that A is a *theorem* of our system, and we write $\vdash A$. We call $A_1, \ldots, A_n$ a *proof sequence for* A.

Example 1: $\vdash (p_1 \rightarrow p_1)$

Analysis Here is a proof sequence that establishes $\vdash (p_1 \rightarrow p_1)$.

$p_1 \rightarrow ((p_1 \rightarrow p_1) \rightarrow p_1)$
$p_1 \rightarrow (p_1 \rightarrow p_1)$
$p_1 \rightarrow ((p_1 \rightarrow p_1) \rightarrow p_1)) \rightarrow ((p_1 \rightarrow (p_1 \rightarrow p_1)) \rightarrow (p_1 \rightarrow p_1))$
$(p_1 \rightarrow (p_1 \rightarrow p_1)) \rightarrow (p_1 \rightarrow p_1)$
$(p_1 \rightarrow p_1)$

Maybe a computer could follow that, but for you and me I'll set out the steps with explanations:

(1) $p_1 \rightarrow ((p_1 \rightarrow p_1) \rightarrow p_1)$
       an instance of axiom scheme 2 taking p_1 for B and $(p_1 \rightarrow p_1)$ for A
(2) $p_1 \rightarrow (p_1 \rightarrow p_1)$
       an instance of axiom scheme 2 taking p_1 for B and p_1 for A
(3) $(p_1 \rightarrow ((p_1 \rightarrow p_1) \rightarrow p_1)) \rightarrow ((p_1 \rightarrow (p_1 \rightarrow p_1)) \rightarrow (p_1 \rightarrow p_1))$
       an instance of axiom scheme 4 taking $(p_1 \rightarrow p_1)$ for B,
       p_1 for A, and p_1 for C
(4) $(p_1 \rightarrow (p_1 \rightarrow p_1)) \rightarrow (p_1 \rightarrow p_1)$ by *modus ponens* using (1) and (3)
(5) $(p_1 \rightarrow p_1)$ by *modus ponens* using (2) and (4)

Example 2: $\vdash (((p_1) \rightarrow (p_1)) \vee (p_7))$

Analysis Here is an (annotated) proof sequence:

(1) $(((p_1) \rightarrow (p_1)) \rightarrow (((p_1) \rightarrow (p_1)) \rightarrow (((p_1) \rightarrow (p_1)) \vee (p_7))))$
       an instance of Axiom 8
(2) $(p_1 \rightarrow p_1)$
       (insert the proof from Example 1)
(3) $(((p_1) \rightarrow (p_1)) \rightarrow (((p_1) \rightarrow (p_1)) \vee (p_7)))$
       modus ponens using (1) and (2)
(4) $((p_1) \rightarrow (p_1)) \vee (p_7))$
       modus ponens using (2) and (3)

As in the last example, we can use any theorem as if it were an axiom in a proof sequence, for we know there is a proof sequence for it that we can insert.

Perhaps when you started you thought that all logic is just symbol pushing. That's certainly not the case with the semantic approach, where we developed the symbols through assumptions about meaning and the world. But proving really is just manipulating the symbols to get a "result". So why bother setting up a formal analysis of proving?

Originally propositional logic was developed entirely syntactically. Only later were truth-tables and models set out to justify the syntactic approach. And the syntactic approach does need justification. Why these axioms? Why this rule?

We justify the axioms by noting that each (instance of an) axiom (scheme) is a tautology (as you can verify). Without that justification, the early developers of propositional logic relied on their "intuition". Then they were confronted with instances of the axiom scheme $B \to (A \to B)$ that many said were nonsense or false, such as "$2 + 2 = 4 \to$ (the moon is made of green cheese $\to 2 + 2 = 4$)". To deny that this example is paradoxical, they had to talk in terms of meaning, essentially giving a truth-table analysis of the connectives, claiming that the "laws of logic" don't and shouldn't take account of relevance or subject matter.

We can justify the rule by noting that if the premises of it are true, then the conclusion is true. So, since the axioms are tautologies, every theorem is a tautology.

But that's not enough. We need that every tautology can be derived as a theorem. It is to the end of making a clear proof of that in Appendix 4 that I've chosen these particular axioms and rule.

But still, why bother with formalizing proving? Those who prefer the syntactic approach, and those who originated propositional logic, said that it's clearer for setting out our assumptions about how to reason well. Here are some simple forms of propositions we can all agree are basic. No talk of meaning and the world. The axiom system implicitly defines the meaning of the formal connectives. Yes, implicitly, but that depends on our having agreed to, or needing to agree to, some assumptions about meaning, truth, and the world.

Still, it is useful to see how we can get all tautologies by starting from some simple ones. We have to simplify to get a system, which in some way does make clearer the relations of the formal connectives. And, as we saw in Chapter 3, deciding whether a wff with forty-seven propositional variables is a tautology is a massive undertaking, while here if it happens to be an instance of one of the axioms or theorems, we're done.

And finally, we want to understand the process of proving. We've proved theorems about the semantic presentation of classical propositional logic, but what logic were we using for that? What are the rules for a proof? By formalizing how to prove, we can better understand and have a standard. We'll need this when we extend our logic in the next section to take account of the internal structure of atomic propositions, for then there is no semantic analysis we can use to decide whether a wff is a tautology, as you can see in Walter Carnielli's and my book *Computability*. So showing how to get tautologies via proving is important. Here we can see it in the simplest form.

Exercises
1. Why bother axiomatizing classical propositional logic?
2. What does it mean to say that there is a proof of A from an axiom system?
3. a. Show that each instance of each scheme of the axiom system for classical propositional logic is valid.
   b. Show that every theorem of the axiom system for classical propositional logic is true in every model.
4. Show that the following are theorems of our axiom system by giving a derivation of each.
   a. $C \to (D \to D)$
   b. $C \to \neg\neg C$

B. Deriving Consequences from Hypotheses

In science and mathematics, even in economics, we start with hypotheses, like "Given any two points, there is a third that lies between them" or "$E = mc^2$". Then we derive consequences from them. We do so by using our (usually implicit) logic. Here we have set out the proof methods, so we can say exactly what we mean by deriving consequences from hypotheses.

Proof from hypotheses A *proof of* B *from a collection of propositions* Σ is a sequence $A_1, \ldots, A_n$, where A_n is B and each A_i is either an axiom, or is in Σ, or is derived from some of the preceding A_j's by one of the rules. In that case, B is a *syntactic consequence of* Σ, and we write $\Sigma \vdash B$.

We sometimes read $\Sigma \vdash B$ as B *is deducible from* Σ, or read $\vdash$ as "perp". We write $A \vdash B$ to mean that B is deducible from just the one wff A, and we write $\Sigma, A \vdash B$ to mean that B is deducible from Σ and A together. We write "$\Sigma \nvdash B$" for "there is not a proof of B from Σ".

Example 3 $\{p_2, \neg p_2\} \vdash p_0$

Analysis
(1) $\neg p_2 \to (p_2 \to p_0)$ an instance of axiom scheme 1
(2) $\neg p_2$ premise
(3) $(p_2 \to p_0)$ by *modus ponens* using (1) and (2)
(4) p_2 premise
(5) p_0 by *modus ponens* using (3) and (4)

Typically, we give schemes of formal proofs, showing for example, $\vdash (A \to A)$ and $\{A, \neg A\} \vdash B$.

Let's be clear: we have not defined the notions of proof and consequence. Relative to this axiom system we have given a definition of $\vdash$, which we read informally as "is a theorem" or "has as a consequence" or "proves". I have to assume you already have an idea of what it means to prove something, and it's that we're formalizing, since we need to use the informal notion to prove theorems about the axiom system. We prove those (informal) theorems in the language of this book, the *metalanguage*, which is English supplemented with technical notions. Here is an example.

***Theorem** Basic properties of the syntactic consequence relation*
  1. $A \vdash A$.
  2. If $\vdash A$, then $\Gamma \vdash A$.
  3. If $A \in \Gamma$, then $\Gamma \vdash A$.
  4. If $\Gamma \vdash A$ and $\Gamma \subseteq \Delta$, then $\Delta \vdash A$.
  5. *Transitivity* If $\Gamma \vdash A$ and $A \vdash B$, then $\Gamma \vdash B$.
  6. *The Cut Rule* If $\Gamma \vdash A$ and $\Delta \cup \{A\} \vdash B$, then $\Gamma \cup \Delta \vdash B$.
  7. If $\Gamma \cup \{A_1, \ldots, A_n\} \vdash B$ and $\Gamma \vdash A_i$ for $i = 1, \ldots, n$, then $\Gamma \vdash B$.

Proof (1) This is immediate by definition.

(2) Any proof sequence for A as a theorem will do to show A is a consequence of Γ.

(3) This is immediate by definition.

(4) Any proof sequence from Γ for A is also a proof sequence from Δ, since every wff in Γ is also in Δ.

(5) Take a proof sequence $A_1, \ldots, A_n$ for A from Γ. Take a proof sequence $B_1, \ldots, B_n$ for B from A. Then $A_1, \ldots, A_n, B_1, \ldots, B_n$ is a proof sequence for B from Γ.

(6) As for part (5) except $B_1, \ldots, B_n$ is a proof sequence for B from $\Delta \cup \{A\}$.

(7) Just insert the proof sequences for $A_1, \ldots, A_n$ from Γ into a proof sequence for B from $\Gamma \cup \{A_1, \ldots, A_n\}$.

These are the syntactic versions of Theorem 2 (p. 31) for the semantic consequence relation. We also have a syntactic version of the Semantic Deduction Theorem, but that's harder to prove (see Appendix 4, p. 174).

The Syntactic Deduction Theorem
$\Gamma \vdash A \rightarrow B$ iff $\Gamma, A \vdash B$.

We want the syntactic and semantic consequence relations to be the same. We've noted that we have that in one direction.

The axiom system for classical propositional logic is sound
For any A and any Γ, if $\Gamma \vdash A$, then $\Gamma \vDash A$.

To show that if $\Gamma \vDash A$ then $\Gamma \vdash A$ is hard. There's a proof of it in Appendix 4. Together these yield the following.

48   An Introduction to Formal Logic

The completeness of the axiom system for classical propositional logic
   For any A and any Γ, $\Gamma \vdash A$ iff $\Gamma \vDash A$.

In disciplines like physics or biology a theory is not just the hypotheses we start with but also the consequences of those hypotheses. Some theories are better than others. The worst are those from which we can derive a contradiction. A very simple form of a contradiction is the pair A,¬A, and we saw in Example 3 that for any B, $\{A, \neg A\} \vdash B$. So if we can prove both A and ¬A from a theory, we can prove any proposition, which makes the theory worthless. The best theories are those from which we can derive as much as possible without deriving a contradiction.

Theories Σ is a *theory* iff for any A, if $\Sigma \vdash A$, then A is in Σ.
   The *theory of* Γ is the collection of all those A such that $\Gamma \vdash A$.

Consistent and complete theories   A collection of propositions Σ is *consistent* iff there is no A such that both $\Sigma \vdash A$ and $\Sigma \vdash \neg A$. A collection is *inconsistent* if it is not consistent.

   A collection of propositions Σ is *complete* iff for every A, either $\Sigma \vdash A$ or $\Sigma \vdash \neg A$.

A complete and consistent theory is as full a description as possible of "a way the world could be", relative to our choice of atomic propositions. We have a semantic notion of a way the world could be: a model of the language. So we have the following, which is proved in Appendix 4.

   A collection of wffs Σ is complete and consistent iff
      there is a model M such that Σ is the collection of all wffs true in M.

Exercises
1. We have the formal and semi-formal languages of classical propositional logic. In what language do we prove facts about those? What is that language called?
2. a. What does $\Sigma \vdash A$ mean?
   b. Prove that if Σ is a collection of theorems and $\Sigma \vdash A$, then $\vdash A$.
3. a. What is a theory?
   b. Show that if Γ is a theory, then every axiom is in Γ.
   c. What is a consistent theory?
   d. What is a complete theory?
   e. What is the relation of complete and consistent theories to models?
4. Prove that for every model M, the collection of wffs true in M is a complete and consistent theory.
5. Prove the theorem on properties of the syntactic consequence relation (p. 47).

Key Words proof metalanguage
axiom The Syntactic Deduction Theorem
axiom scheme soundness of the axiom system
theorem completeness of the axiom system
proof sequence theory
proof from hypotheses consistent theory
syntactic consequence complete theory

Summary of Chapters 1 to 6

Logic is meant as a guide for establishing what's true and what's false, (propositions) and what propositions follow from others (inferences) based on certain assumptions about the forms of propositions and about meaning and truth.

To simplify our investigations, we focussed on the forms of propositions as built from other propositions using connectives we abstracted from "and", "or", "not", and "if . . . then". To make precise those forms, we set up a formal language. Then the propositions we can analyze are those of a semi-formal language, which is what we get when we replace the variables in the formal language with ordinary-language propositions.

Then we adopted the classical abstraction to ignore all properties of propositions except their form as built from other propositions and their truth-value. Given certain simplifying assumptions about how the truth-value of a compound proposition depends on the truth-values of its parts, we settled on the classical truth-tables as our interpretations of the formal versions of the connectives. A model, then, is a semi-formal language with truth-values assigned to the atomic propositions and the truth-values of the compound propositions derived from the truth-values of the atomic propositions in it by the classical truth-tables.

A semi-formal proposition that is evaluated as true in every model is a tautology. Relative to our assumptions about form and meaning, this formalizes the idea that a proposition is true due to its form. An inference of the semi-formal language is valid just in case every model in which the premises are true, the conclusion is, too. This formalizes the idea that an inference is valid due to its form, at least relative to our assumptions about form and meaning. It depends on understanding a way the world could be as a model, relative to our assumptions about what counts as form and truth.

Then we saw a syntactic characterization of tautology and inference, relative to the forms of propositions we're considering. We gave an analysis of proving and set out an axiom system from which we could derive exactly the tautologies as theorems and which we could use to characterize valid inferences.

Along the way we saw examples of how to use our logic to formalize ordinary-language propositions and inferences. And we saw limitations on what we could formalize.

Further reading
You can see a fuller development of classical propositional logic along with a history in my *Propositional Logics*. There you can also find formal logics that take account of semantic values other than just truth-value, such as subject matter, or that assume there are more than two truth-values. In my *Time and Space in Formal Logic*, you can see formal propositional logics that takes account of the time as a semantic value.

7   Reasoning about Things

In classical propositional logic we have atomic propositions and ways to join them with the four connectives. That plus our assumptions that we're concerned only with truth-values and the division of form and content gave us a simple logic we can use to evaluate reasoning. We can make clear exactly why the forms of valid inferences we saw in Chapter 2 are valid and why the invalid ones are invalid. We can evaluate any inference as long as all that's important is its form relative to the four connectives. But that leaves much we can't evaluate with classical propositional logic.

Example 1:   *All dogs bark. Ralph is a dog. So Ralph barks.*
Analysis   This is valid, but the form of it in classical propositional logic is just p, q, therefore r, which not a scheme of valid inferences.

Example 2:   *All good teachers give fair exams. Prof. Zzzyzzx gives fair exams.*
So *Prof. Zzzyzzx is a good teacher.*
Analysis   This may seem valid, but it's not: the premises could be true, yet Prof. Zzzyzzx could be a terrible teacher and give fair exams from an instructor's manual.

Example 3:   *Some dogs like cats. Some cats like dogs. So some dogs and cats like each other.*
Analysis   This seems valid, too. But it's not. It could be that all the dogs that like cats are scorned by the cats as too wimpy.

Example 4:   *Snow is white. What's white reflects light. So snow reflects light.*
Analysis   This seems valid. But then what about:
> Snow is white.
> White is the color that results from reflecting all wave lengths of light.
> So snow is the color that results from reflecting all wave lengths of light.

That's not valid. What's wrong? Propositional logic is no help.

We could proceed piecemeal, looking at one inference and another, trying to see some patterns as we did in Chapter 2. But there are so many forms to consider once we begin looking at the structure of atomic propositions we need some guide, some general approach to begin our studies.
   The most basic grammatical form of English sentences is that of subject and predicate. The subject tells us what we're talking about; the predicate describes the subject. So the subject of "Prof. Zzzyzzx gives fair exams" is the phrase "Prof. Zzzyzzx", and the predicate is "gives fair exams". The subject of "Snow is white" is the word "snow" and the predicate is "is white". But what is the subject of "Some dogs like cats"? Shall we say it's "some dogs"?

Much of our talk in English is about things: dogs, cats, tables, Ralph, rocks, flies. We have lots of words for kinds of things and lots of names for particular things. We organize much of our experience through our language in terms of things. But not all, for we also talk about water and mud, and the burning of a flame in a fireplace, and the push of the wind. Still, the grammar of English is primarily organized around talk of things. So let's focus on the internal structure of atomic propositions in terms of things.

Things, the World, and Propositions The world is made up at least in part of things. The only propositions we are interested in are about things.

What do we mean by "thing"? We seem to be able to agree that rocks, people, dogs, tables, chairs, and trees are things. What we consider most basic about them is not what they're made of or whether we happen to be looking at them but that they are individuals: this rock, that person, this dog, that tree. A thing, whatever it is, is individual and distinct from all else in the world. Yes, a thing may be composed of other things or masses, but what makes it a thing is that it is a whole, a distinct individual. How odd that sounds, for we seem to be saying over and over what we have no way to say except by saying "a thing" or "an individual".

Since each thing is distinct from all else, each is in some way distinguishable from all else. I can distinguish this chair from anything else. Or at least I think I can, since it's old and there's not likely to be another that looks like it. I can distinguish my computer from all other things in the world, for though it looks like other ones, it contains data others don't. I can distinguish this grain of sand from all other grains on the beach, so long as it's in my hand. I can distinguish this pig from all other pigs in Denmark, though that's pretty theoretical since I suspect that by the time I saw the 4,319th pig, I'd be confused. But we can agree that in principle I can distinguish it; certainly I can distinguish it from my dog. But what about a star that someone says must be out there though no one's seen it? Perhaps we can distinguish it by the effects it has on other stars. What about the number 7? How do I distinguish it from 43 if I can't touch, see, smell, taste, or hear it?

What we mean by saying that we can distinguish each thing from all others will be determined in part by the kind of things we are talking about. Equally, it will determine what we consider to be a thing. There is no fixed answer to our question of what we mean by "distinguishable" that we can agree on for all things. Nonetheless, this is where we'll start, refining and comparing our notions of thing and distinguishability as we proceed in our work.

The Distinguishability of Things Each thing is in some way distinguishable from all other things. It can be identified and re-identified.

Chapter 7 Reasoning about Things 53

These are large assumptions. Are they true? I am at a loss to know what that means. They describe ways we encounter the world through our language, ways that lie behind much of our talk and reasoning. That is enough for us to proceed.

Aside: *Grammar as the basis of validity*
Some say that we needn't worry about the nature of things or make such grand assumptions. We can stop with a grammatical analysis. We can make a list of parts of language:

   names
   words we can preface with "the", "a", or "an"
   pronouns such as "this", "that", and "it"

"Thing", they say, is a neutral word, like "it". It has no meaning in itself but simply stands ready to be replaced by a word from a list such as this one. Logic, they argue, is independent of what the words mean, though it does assume they are meaningful. In propositional logic we did not ask what or how atomic propositions meant. We only agreed that they were meaningful in the sense that each was (viewed as being) true or false. From $A \wedge B$ we could conclude B without concerning ourselves about the relation of each proposition to "the world".

But when we consider the internal structure of atomic propositions, this syntactic approach will not do. The word "thing" is not a marker of a grammatical category. The following deduction is valid:

   Peter is in a car.
   Therefore, there exists a car.

The reason: a car is a thing, it exists. But we cannot claim the following are valid, though they share the same grammatical form:

   Peter is in a hurry.
   Therefore, there exists a hurry.

   Peter is in a huff.
   Therefore, there exists a huff.

We say "Peter did it for the sake of Dick" and "Tom left Dick in the lurch", but we do not think that "sake" and "lurch" stand for things, and we will not reason with them as if they do.

Those who want to rely on only a grammatical analysis disagree. They say that we can reason about lurches and sakes just as well as about cats and dogs. All we need is that the words are meaningful (but not necessarily that *we* know what they mean) and that they play the correct grammatical role. So whatever reasons we use to justify that (a) is valid will suffice to justify the validity of: (b)

(a)   All dogs bark.
      Ralph is a dog.
      Therefore, Ralph barks.

(b)   All sakes bark.
      Ralph is a sake.
      Therefore, Ralph barks.

But the reasoning in (b) has only the pattern of validity without being valid. We are not reasoning about sakes but only with the word "sake" acting as a marker for appropriate nouns. The apparent validity of (b) is parasitic on our intuition for inferences about things.

Key Words   Things, the World, and Propositions
　　　　　　　The Distinguishability of Things

8   The Grammar of Things

    A.   Names and Predicates 55
    B.   Propositional Connectives 56
    C.   Variables and Quantifiers 57
    D.   The Grammar of Predicate Logic 59
    Exercises 59

B. Names and Predicates

If propositions are about individual things, then words that pick out or purport to pick out specific individual things are of special importance. Those are names, for example, "Socrates", "Marie Antoinette", "Arf", "Santa Claus".

    An example of the simplest kind of proposition that involves a name is:

    Ralph is sitting.

We have a name "Ralph" and what grammarians say is a predicate, "is sitting". There's no further for our analysis to go.

    Here is a proposition with two names:

(1)   Dick loves Spot.

We can view this as a name, "Dick", and a predicate, "loves Spot". Or we can view (1) as being about Spot, and hence it's a name, "Spot", and what's asserted about him, "Dick loves". Or we can view (1) as about both Dick and Spot, composed of two names and a predicate, "loves". This strains the usual reading of the word "predicate" you learned in school, but we can say that a predicate arises by deleting either or both of the names in (1).

    Here's a proposition that uses three names:

    Beulah and Fido are the parents of Spot.

There are seven ways to parse this proposition depending on whether we consider it to be about one, two, or all three of the objects named in it. For instance, we can view it as a proposition about Beulah and Spot, and hence the predicate that arises is "— and Fido are the parents of —". I've used blanks where I deleted names to make it clear where the names go. In any proposition that contains names, we can obtain a predicate by deleting one or more of the names.

Names   A *name* is a word that is meant to pick out a specific individual thing.

Predicates   A *predicate* is any incomplete (English) phrase with specified gaps such that when the gaps are filled with names the phrase becomes a proposition.

    These definitions strongly reflect our assumption that the world is made up of things. If we had begun our analysis by assuming that the world is made up of processes, we might parse propositions into verbs and what's left over when the verbs are deleted.

This definition seems to say that no matter what names fill in the gaps in a predicate, a proposition results. But if "— loves —" is a predicate, what are we to make of "7 loves Juney"? We have two choices. We may say that when we took "— loves —" to be a predicate, we weren't thinking of names of all things as being suitable to fill in the blanks but only, say, names of people and animals. That is, we can restrict what kinds of things we're talking about. Alternatively, we can assume that there is a homogeneity to all things in that whatever can be asserted about one thing can be asserted about any other, though of course the assertion may be false. In that case, "7 loves Juney" would be false. We'll look at both these choices in what follows.

To reduce ambiguity, let's require that each name-place of a predicate has a separate dash associated with it and that for any particular predicate the number of dashes won't vary. We don't want "— are the brothers of —" to be a predicate if by that we mean the blanks can be filled with as many names as you wish. If a predicate has one blank it is *unary*, if two it is *binary*, if three it is *ternary*, if forty-seven it is forty-seven-ary. The number of blanks in a predicate is its *arity*.* Predicates other than unary ones are called *relations*. When it's clear where and how many blanks occur in a predicate, we may informally leave the dashes off, saying that "barks" or "is green" or "is the mother of" are predicates.

So "— barks" is a predicate and "Ralph" is a name. When we put them together, we get "Ralph barks". But in that expression we've lost the distinction of the predicate and the name. So let's write instead:

(— barks) (Ralph)

Predicate and name are clearly distinguished. Similarly, for (1) we'll write:

(— loves —) (Dick, Spot)

This shows that the name "Dick" is meant to fill in the first blank, and the name "Spot" is meant to fill in the second blank. We're using blanks, and parentheses, and commas for punctuation here.

This definition of "predicate" recognizes two grammatical categories: propositions and names. What other grammatical categories shall we recognize?

B. Propositional Connectives

We continue to take propositions as fundamental. So we can continue to use the ways of forming new propositions from others as in propositional logic, taking $\wedge, \vee, \rightarrow, \neg$ for formalizations of "and", "or", "if . . . then . . .", and "not".

An atomic proposition in propositional logic is one whose internal structure is not taken into account. For example, "Ralph barks" is atomic. But now we recognize the structure of that proposition as made up of a name and a predicate.

---
* I am not responsible for this barbarism; it comes with the subject.

Nonetheless, let's continue to classify "(— barks) (Ralph)" as *atomic* because it contains no formal symbol of our logic.

C. Variables, and Quantifiers

So far we've considered only propositions about specific named things. But the following is about things, too:

(2)   All dogs bark.

This isn't about any specific dog. There's no name in it, nor predicate. Still, it is about things: dogs, all dogs. It says that each and every dog barks. We have the predicates "— is a dog" and "— barks". What we need is a way to connect them and say "each and every" or simply "all".

The sentence (2) doesn't say that everything is a dog and barks. Rather,

Given any thing, if it is a dog, then it barks

The pronoun "it" is like a name, a temporary name. You're out walking with me and see a fleeting shape ahead and ask what it is. I say "It is a dog", for I don't know a name for it (of if it even has one). Or I can point at some puppet in a store window and say, "It is a dog"—which would be false. But if I want to say that this is a dog and that is a cat, I can't use "it is a dog and it is a cat" because then "it" would have two meanings according to what it's used as a temporary name for. So instead of using just the few pronouns we have in English like "it", "this", and "that", let's make sure we have plenty of words we can use as temporary names:

$x_1, x_2, x_3, \ldots$

We call these *variables*. Informally, I'll use x, y, and z as variables for x_1, x_2, and x_3.

Now we can formalize "if it is a dog, then it barks":

(— is a dog) (x) → (— barks) (x)

How shall we say "all"? We could write:

For all x, (— is a dog) (x) → (— barks) (x)

And that's true. After all, if "x" is used as a temporary name of the lamp on my table, then "(— is a dog) (x) → (— barks) (x)" is true, since the antecedent is false. That's why we evaluate the conditional to be true when it has a false antecedent.

Talking about all dogs, all cats, all whole numbers, all tables, is common enough and central enough in our reasoning to formalize. Let's use the symbol $\forall$ to formalize "all". Then we can formalize "All dogs bark" as:

$\forall x,$ (— is a dog) (x) → (— barks) (x)

Using parentheses instead of commas to mark phrases, we'll have:

$\forall x\,(\,(-\text{ is a dog})\,(x) \to (-\text{ barks})\,(x)\,)$

Along with talk of all comes talk of some, as in:

Some cat is feral.

That is, there is some thing such that it is a cat and it likes to swim. We can replace that pronoun "it" with a variable:

For some thing x, $(\,(-\text{ is a cat})\,(x) \land (-\text{ is feral})\,(x)\,)$

Our use of "some" is important enough in our reasoning to formalize. Let's use the symbol $\exists$ to formalize "some". So we'll write:

(3) $\exists x\,(\,(-\text{ is a cat})\,(x) \land (-\text{ is feral})\,(x)\,)$

In what we're doing here, I have to assume you understand the words and phrases "there is a", "for all", "for some", "there exists", just as I had to assume you understood "and" when we introduced the symbol $\land$. We are replacing these phrases with formal symbols to which we'll give precise meanings based on our understanding of English. Then we'll be able to claim that their use in formulas yields propositions.

We call $\forall$ the *universal quantifier* ; reading it out loud we'll say "for all". We call $\exists$ the *existential quantifier* ; reading it out loud we'll say "there exists". The use of a quantifier to turn "$(\,(-\text{ is a cat})\,(x) \land (-\text{ likes to swim})\,(x)\,)$" into the proposition (3) is called a *quantification*.

Note that now a proposition that is not atomic need have no part that is atomic: in (3), the part "$(-\text{ is a cat})\,(x)$" is not a proposition but becomes one only when we say what thing x is meant to stand for or we add a quantifier to show we're talking about all or some things as in "$\exists x\,(-\text{ is a cat})\,(x)$".

Example 1: $(-\text{ is a parent of }-)\,(x, y)$

Analysis This isn't a proposition. It isn't true nor is it false unless we know what x and y are meant to stand for, just as "this is a parent of that" is not true or false unless we know what "this" and "that" are meant to pick out. But if we attach quantifiers to the example, we can get a proposition, for example:

$\forall x\,\exists y\,(\,(-\text{ is a parent of }-)\,(x, y)\,)$

This is true of people. Here are other ways we could convert the example into a proposition using quantifiers:

$\forall x\,\forall y\,(\,(-\text{ is a parent of }-)\,(x, y)\,)$

$\exists x\,\forall y\,(\,(-\text{ is a parent of }-)\,(x, y)\,)$

$\exists x\,\exists y\,(\,(-\text{ is a parent of }-)\,(x, y)\,)$

Of these, only the last is true of all human beings who have ever lived.

Example 2: $\exists x\,((-\text{ is a parent of }-)(x, y))$

Analysis This, too, isn't true or false until we say what y is meant to stand for, or we use a quantification for y. When we define a formal language, we'll have to pay attention to which variables in a wff are "free", meaning not affected by a quantifier.

D. The Grammar of Predicate Logic

The parts of speech we've considered in our analysis of the internal structure of atomic propositions are:

  propositions
  names
  predicates
  propositional connectives
  quantifiers
  punctuation

Predicates are derivative from our choice of what atomic propositions and names we take. Atomic propositions, though primitive in that they contain no formal symbols of our logic, are no longer structureless: they are parsed as a name or names and what's left over when the name or names are deleted. We use blanks and parentheses to make clear how we're combining a predicate with a name or names, and we use parentheses to mark phrases.

Compound propositions are formed from predicates and names by using propositional connectives and, with the use of variables as temporary names, quantifiers. To formalize the use of ordinary English quantifiers we use variables and formal quantifiers, to which we'll give interpretations once we've agreed on the exact forms we're considering.

This list was motivated by Things, the World, and Propositions. But it doesn't exhaust the ways we can look at the structure of propositions that reflect that assumption. Nonetheless, it will provide us with enough structure to analyze many kinds of inferences we could not previously analyze. The formal versions of these parts of speech will comprise the syntax of *predicate logic*.

Exercises

1. What two assumptions about the world and language do we make to start predicate logic?
2. What is a name?
3. For each of the following propositions list the predicate(s) and name(s) and comment on your choices relative to *Things, the World, and Propositions*.
   a. Dick went to see Tom.
   b. Ralph is a dog or he's a puppet.
   c. Ralph hit Juney .
   d. Ralph read *An Introduction to Formal Logic*.

e. 8 is less than 9
f. Paris is in France.
g. Horatio was sick or tired.
h. Juney isn't a puppet.
i. The set of natural numbers contains 7.
j. Green is pleasanter than gray.

4. a. Is "7 loves Ralph" a proposition?
    b. Is "Ralph is divisible by Juney" a proposition?
5. a. What is the universal quantifier? What is it meant to formalize?
    b. What is the existential quantifier? What is it meant to formalize?
6. What use of quantifiers would make the following into true propositions?
    a. $(-\text{ is a dog})(x) \wedge (-\text{ is a woman})(y) \wedge (-\text{ loves }-)(x, y)$
    b. $(-\text{ is the biological mother of }-)(x, y)$

Key Words

names	relation
predicates	temporary name
grammar	variable
unary predicate	quantifier
binary predicate	universal quantifier
arity	existential quantifier

9 A Formal Language for Predicate Logic

A. The Formal Language 61
B. Realizations: Semi-Formal English 63
Exercises 66

A. The Formal Language

We have no list of all predicates and names nor any method for generating them, for English is not a fixed, static language. We might use $P_0, P_1, P_2, \ldots$ to stand for predicates, but that wouldn't account for the arity of the predicate. So we'll use $P_0^1, P_0^2, P_0^3, \ldots, P_1^1, P_1^2, P_1^3, \ldots$ as *predicate symbols* where the subscript tells which one in the list it is and the superscript tells the arity. We can use $c_0, c_1, c_2, \ldots$ as *name symbols* to stand for names. Collectively the name symbols and variables are *terms*. The ellipses "..." are to indicate that the list goes on indefinitely. We need not take more than, say, 47 predicate, 12 name symbols, and 6 variables. But it's simpler to assume that there are always enough when we need them. And rather than taking $P_2^3(x_6 \; x_1 \; x_{14})$ as a formula, let's use commas between the variables to make it easier to read, as in $P_2^3(x_6, x_1, x_{14})$.

To talk about the formal language we'll need *metavariables*. We can use $t, u, v, t_0, t_1, \ldots$ to stand for any term; $x, y, z, w, y_0, y_1, \ldots$ to stand for variables; and $A, B, C, A_0, A_1, \ldots, B_0, B_1, \ldots$ to stand for formulas. If we use $i, j, k, m, n, i_0, i_1, \ldots$ to stand for counting number, then, for example, P_i^n can stand for any predicate symbol. I'll also use P and Q to stand for a predicate symbol, with the arity clear from context.

The formal language of predicate logic

Vocabulary predicate symbols P_i^n for $n \geq 1$ and $i \geq 0$, where n is the arity
   name symbols $c_0, c_1, \ldots$ ⎫
   variables $x_0, x_1, \ldots$ ⎬ terms
                                                  ⎭
   propositional connectives ¬, →, ∧, ∨
   quantifiers ∀, ∃

Punctuation parentheses ( )
   comma ,

Grammar (well-formed formulas — *wffs*)
  i. If P is a k-ary predicate symbol and $t_1, \ldots, t_k$ are terms, then $(P(t_1, \ldots, t_k))$ is a wff of *length* 1.
   If t_i is a variable, it is *free* in the wff.

ii. If A is a wff of length n, then $(\neg A)$ is a wff of length $n + 1$.
An occurrence of a variable in $(\neg A)$ is free iff it is free in A.

iii. If A and B are wffs, and the maximum of the lengths of A and B is n, then each of $(A \rightarrow B)$ and $(A \wedge B)$ and $(A \vee B)$ is a wff of length $n + 1$.
An occurrence of a variable in $(A \rightarrow B)$ is free iff the corresponding occurrence of the variable in A or in B is free, and similarly for $(A \wedge B)$ and $(A \vee B)$.

iv. If A is a wff of length n and some occurrence of x is free in A, then each of $(\forall x A)$ and $(\exists x A)$ is a wff of length $n + 1$.
An occurrence of a variable in either $(\forall x A)$ or $(\exists x A)$ is free iff the variable is not x and the corresponding occurrence in A is free.

v. A concatenation of symbols is a *wff* iff it is a wff of length n for some $n \geq 1$.

Wffs of length 1 are *atomic*. All other wffs are *compound*.

In $(\forall x A)$ the initial $\forall x$ has *scope* A and *binds* each free occurrence of x in A. The occurrence of x immediately after $\forall$ is also *bound*. Similarly for $(\exists x A)$.

A wff is *closed* if there is no occurrence of a variable free in it; otherwise it is *open*.

Theorem 1   The unique readability of wffs of the language of predicate logic
There is one and only one way to parse each wff of the formal language.

Proof   To each symbol α of the formal language assign an integer $\lambda(\alpha)$ according to the following chart:

$\neg$	$\rightarrow$	$\wedge$	$\vee$	$-$	,	(	)	$\forall$	$\exists$	x_i	c_i	P_i^n
0	0	0	0	0	0	-1	1	-1	-1	1	1	$-n$

To the concatenation of symbols $\alpha_1 \alpha_2 \cdots \alpha_n$ assign the number:

$$\lambda(\alpha_1 \alpha_2 \cdots \alpha_n) = \lambda(\alpha_1) + \lambda(\alpha_2) + \cdots + \lambda(\alpha_n)$$

We proceed by induction on the number of occurrences of $\forall, \exists, \neg, \rightarrow, \wedge, \vee$ in A to show that for any wff A, $\lambda(A) = 0$.

If there are no occurrences, then A is atomic. That is, for some $i \geq 0$, $n \geq 1$, A is $(P_i^n(-, \ldots, -)(u_1, \ldots, u_n))$, where there are n blanks and $n - 1$ commas. I'll let you calculate that $\lambda(A) = 0$.

The inductive stage of the proof, and then the proof that initial segments and final segments of wffs have value $\neq 0$, and then that there is only one way to parse each wff, are done almost the same as for propositional language (p. 16). ∎

Example 1: $(P_1^2(x_7, c_{16}))$
Analysis   This is a well-formed formula. The single occurrence of x_7 is free. The formula is open and atomic. It's length is 1.

Example 2: $(\exists x_7 (P_1^2(x_7, c_{16})))$
Analysis This is a well-formed formula. The occurrences of x_7 are bound by the quantifier. The formula is closed and compound. Its length is 2.

Example 3: $((\forall x_7 (P_0^1(x_7))) \to (P_1^2(x_7, c_{16})))$
Analysis This is a well-formed formula. The scope of $\forall x_7$ is $(P_0^1(x_7))$. The occurrence of x_7 in $P_0^1(x_7)$ is bound by the quantifier; the occurrence of x_7 in $P_1^2(x_7, c_{16})$ is free. The formula is open and compound. Its length is 3.

$((\forall x_7 (P_0^1(x_7))) \to (P_1^2(x_7, c_{16})))$
                                    1
              2
                      3

We name a formal language by its vocabulary. In this case it is:
$L(\neg, \to, \wedge, \vee, \forall, \exists, x_0, x_1, \ldots, P_0^1, P_0^2, P_0^3, \ldots, P_1^1, P_1^2, P_1^3, \ldots, c_0, c_1, \ldots)$.
We'll abbreviate that as:

$L(\neg, \to, \wedge, \vee, \forall, \exists, P_0, P_1, \ldots, c_0, c_1, \ldots)$

We need more conventions for informally deleting parentheses:
- The parentheses around atomic wffs and the outer parentheses around the entire wff can be deleted.
- Parentheses between successive quantifiers at the beginning of a wff may be deleted.
- $\neg$ binds more strongly than $\wedge$ and $\vee$, which bind more strongly than $\to$. And $\forall x, \exists x$ bind more strongly than any connective.
- Square brackets [ ] may be used in place of parentheses.

Example 4: $(\forall x_1 (\exists x_2 (\forall x_3 (P_1^3(x_4, x_2, x_3)))))$
Analysis We can abbreviate this as: $\forall x_1 \exists x_2 \forall x_3 P_1(x_4, x_2, x_3)$

B. Realizations: Semi-Formal English
Just as with classical propositional logic, we want to use our formal language as a guide to analyzing ordinary language propositions and inferences.

Realizations and semi-formal languages An ordinary language name or predicate is *simple* if it contains no part that we could formalize as a name, predicate, propositional connective, variable, or quantifier, or a combination of those.
     A *realization* of the formal language is an assignment of simple names to none, some, or all of the name symbols and an assignment of simple predicates to at least one of the predicate symbols. The *realization of a formal wff* is what

we have when we replace the formal symbols in it with the parts of ordinary language that are assigned to them; it is a *semi-formal wff*. The *semi-formal language* for a realization is the collection of realizations of the formal wffs.

We must realize at least one predicate symbol in order to have any semi-formal wffs at all.

Example 5: *A realization*

$$L(\neg, \rightarrow, \wedge, \vee, \forall, \exists, P_0, P_1, \ldots, c_0, c_1, \ldots)$$
$$\downarrow$$
$$L(\neg, \rightarrow, \wedge, \vee, \forall, \exists\, ; \text{``$-$ is a dog''}, \text{``$-$ is a cat''}, \text{``$-$ eats grass''}, \text{``$-$ is a wombat''}, \text{``$-$ is the father of $-$''}; \text{``Ralph''}, \text{``Dusty''}, \text{``Howie''}, \text{``Juney''})$$

Analysis The predicates realize the predicate symbols of the correct arity in order, and the names realize the name symbols in order. So "$-$ eats grass" realizes P_3^1, and "Dusty" realizes c_1. The expressions or formulas of the semi-formal language are the realizations of the formal wffs. For example, "($-$ is a dog) (Ralph)" is an expression of the semi-formal language, realizing $P_0^1(c_0)$: real($P_0^1(c_0)$) = "($-$ is a dog) (Ralph)".

We know that the predicates and names in a presentation of a realization are pieces of language. So we can delete the quotation marks and write the realization of Example 5 as:

$$L(\neg, \rightarrow, \wedge, \vee, \forall, \exists, P_0, P_1, \ldots\ldots, c_0, c_1, \ldots)$$
$$\downarrow$$
$$L(\neg, \rightarrow, \wedge, \vee, \forall, \exists\,;\ -\text{ is a dog},\ -\text{ is a cat},\ -\text{ eats grass},\\ -\text{ is a wombat},\ -\text{ is the father of}; \text{Ralph, Dusty, Howie, Juney})$$

Example 6: The realization of the (abbreviated) formal wff:
$$P_0^1(c_0) \wedge P_0^1(c_3) \rightarrow \neg \exists x_1\, P_1^1(x_1)$$
is (($-$ is a dog) (Ralph) $\wedge$ ($-$ is a dog) (Juney)) $\rightarrow \neg \exists x_1$ ($-$ is a cat) (x_1)

The realization of the formal wff:
$$(\neg P_0^1(c_0)) \rightarrow \forall x_0\, (P_1^1(x_0))$$
is ($-$ is a dog) (Ralph) $\rightarrow (\forall x_0$ ($-$ is a cat) (x_0))

Analysis These closed wffs will be propositions only when we've settled on how we will interpret the formal symbols.

The terminology we've adopted to discuss the formal language can be carried over to apply to semi-formal wffs.

Example 7: $\forall x_1$ (($-$ is a dog) (x_1) $\wedge$ ($-$ is bigger than $-$) (x_1, Ralph)
$\rightarrow \exists x_2$ ($-$ is smaller than $-$) (x_2, x_1))

Analysis The second occurrence of x_2 in this abbreviated wff is bound by the quantifier $\exists x_2$ and the predicate "$-$ is bigger than $-$" appears in the semi-formal wff.

Categorematic vocabulary, logical vocabulary, and punctuation

The *categorematic* vocabulary of a semi-formal language consists of the predicates that realize the predicate symbols and the names that realize the name symbols.

The *syncategorematic* or *logical* vocabulary of a semi-formal language consists of the formal symbols $\forall, \exists, \neg, \rightarrow, \wedge, \vee, x_0, x_1, \ldots$.

Punctuation is that part of the vocabulary that is not meant to formalize anything but is used only to facilitate reading wffs.

Categorematic parts of the formal language joined by logical vocabulary and punctuation are categorematic.

Example 8: $(-\text{ is green})(x_3)$
Analysis This is not a proposition until we say what x_3 is meant to stand for.

Example 9: $(-\text{ loves }-)(x_2, \text{Juney}) \rightarrow (-\text{ loves }-)(\text{Arf}, \text{Juney})$
Analysis This is not a proposition until we say what x_2 is meant to stand for.

Example 10: $\forall x_1 \,(-\text{ is a dog})(x_1) \rightarrow (-\text{ barks})(x_2)$
Analysis This, too, cannot be viewed as a proposition until we say what x_2 is meant to stand for.

Only closed formulas of the semi-formal language can be propositions.

Example 11: $\forall x_1 \,(-\text{ is a dog})(x_1) \rightarrow \exists x_2 \,(-\text{ barks})(x_2)$
Analysis Once we agree on how we'll understand the quantifiers, this will be a proposition. There is no variable free in it.

Example 12: $(-\text{ is less than }-)(x_1, x_2) \rightarrow \neg(-\text{ is less than }-)(x_2, x_1)$
Analysis If understood to be about the counting numbers, most mathematicians and many logicians would take this to be a proposition, viewing it as about all counting numbers. They understand the example as meaning the same as:

$\forall x_1 \forall x_2 \,[\,(-\text{ is less than }-)(x_1, x_2) \rightarrow \neg(-\text{ is less than }-)(x_2, x_1)\,]$

But this is worse than confusing. We expect a proposition and its negation to have opposite truth-values. Yet both "$(-\text{ is even})(x_1)$" and "$\neg((-\text{ is even})(x_1))$" would be false of the counting numbers since the latter is to be understood as "$\forall x \neg((-\text{ is even})(x_1))$". We'll always be explicit when we intend all the variables to be universally quantified.

The universal closure of a wff Let $x_{i_1}, \ldots, x_{i_n}$ be a list of all the variables that occur free in A in *alphabetical order*, that is, $i_1 < \cdots < i_n$. The *universal closure* of A is

$$\forall \ldots A \equiv_{\text{Def}} \forall x_{i_1} \cdots \forall x_{i_n} A.$$

Example 13: The universal closure of
$$((x_9 \text{ is a dog}) \wedge (x_2 \text{ is a cat})) \rightarrow (\exists x_1 (x_1 \text{ chases } x_4))$$
is $\quad \forall x_2 \, \forall x_4 \, \forall x_9 \, [(x_9 \text{ is a dog}) \wedge (x_2 \text{ is a cat})) \rightarrow (\exists x_1 (x_1 \text{ chases } x_4))]$

When we establish a realization, we assume the meaningfulness of its parts. We're about to give an explanation of the syncategorematic parts, which will ensure that wffs in which they appear are meaningful, too. We've also defined what it means for a concatenation of symbols to be well-formed. Shall we then assume that every well-formed formula of the semi-formal language is meaningful? What about "(— loves —) (Ralph, 7)"? It's a well-formed formula. But it seems to be just nonsense. Should we say it's a proposition?

If we were to put further restrictions on what formulas of the semi-formal language are meaningful more than to say they are well-formed, then we are embarked on a semantic analysis not taken into account by the assumptions we've adopted. Rather than do that, we'll give a reading of the syncategorematic parts of the language in such a way that we can ascribe meaningfulness to all well-formed formulas, that is, ascribe semantic properties to them, noting only the one grand division between closed formulas, to which truth or falsity can be ascribed, and open formulas, to which those terms cannot apply.

Form and Meaningfulness What is grammatical and meaningful is determined solely by form and what primitive parts of speech are taken as meaningful. In particular, given a semi-formal language, every well-formed formula will be taken as meaningful, and every closed formula will be taken as a proposition.

Exercises
1. Why in a semi-formal language can only closed formulas be propositions?
2. List the categorematic and syncategorematic parts of these abbreviated semi-formal wffs:
    a. $\forall x_1 (((-\text{ is a dog})(\text{Ralph}) \wedge \neg (-\text{ is a cat})(x_1))$
       $\rightarrow (-\text{ likes }-)(\text{Ralph}, x_1))$
    b. $(-\text{ knew }-)(\text{Plato, Socrates}) \rightarrow (-\text{ is dead})(\text{Plato})$
    c. $\forall x_2 ((-\text{ belongs to }-)(x_2, \text{Arf}) \rightarrow (-\text{ was made in America})(x_2))$
    d. $\forall x_6 ((-\text{ belongs to the United Nations})(x_6) \rightarrow (-\text{ has a president})(x_6))$
3. Give an example (abbreviated if you wish) of a formal wff that is:
    a. Atomic, open, and contains at least one name symbol.
    b. A universal wff.
    c. A universal open wff.
    d. A universal quantifier applied to an existential quantifier of a conditional whose antecedent is a universal formula.
    e. A wff that is open but which contains a closed formula.
    f. A compound proposition no part of which is a proposition.

4. For each part below if it is a formal unabbreviated wff state:
   i. Whether it is atomic.
   ii. Whether it is a universal or existential formula or neither).
   a. $P_1^1(x_1)$
   b. $(P_2^2(x_2))$
   c. $(P_2^1(x_1))$
   d. $(\forall x_1 (P_{47}^1(x_2)))$
   e. $(\exists x_2 (P_2^1(x_2)))$
   f. $((P_3^1(x_4)) \to (\neg(P_4^2(x_2,x_1))))$
   g. $\neg P_5^2(x_1,x_2) \to (P_1^2(x_2,x_1) \vee \neg P_1^2(x_2,x_1))$

5. Convert each of the following abbreviations into a semi-formal wff and give its length.
   a. $\forall x_1 (-$ is a dog$)(x_1)$
   b. $\forall x_1 \exists x_2 ((- $ is a dog$)(x_1)) \to (- $ is a cat$)(x_2) \wedge (- $ hates $-)(x_1, x_2))$

6. For each of the following abbreviated wffs:
   i. Identify the scope of each quantifier.
   ii. State which occurrences of variables are free and which are bound.
   iii. State whether the wff is a closed or open, as well as whether it is atomic.
   iv. Give the universal closure of the wff.
   a. $P_1(c_8)$
   b. $P_8(x_1)$
   c. $(\exists x_1 P_1(x_1) \vee \neg P_2(c_3))$
   d. $\forall x_1 (\neg P_2(x_1) \vee P_1(x_3)) \to \exists x_3 P_4(x_1,x_3)$
   e. $\forall x_2 P_1(x_1,x_2,x_3) \to \exists x_2 \exists x_3 P_1(x_1,x_2,x_3)$
   f. $P_1(x_1) \vee \neg P_1(x_1)$
   g. $\forall x_1 P_1(x_1) \wedge \neg \exists x_2 P_2(x_2)$

7. For each of the following abbreviated semi-formal wffs:
   i. Identify the scope of each quantifier.
   ii. State which occurrences of variables are free and which are bound.
   iii. State whether the wff is a sentence or an open expression, as well as whether it is atomic.
   iv. Give the universal closure of the wff.
   a. $(- $ is an uncle of $-)(x_1, x_2) \vee \exists x_1 (- $ is a woman$)(x_1)$
   b. $(- $ is an uncle$)($Ralph$) \vee \exists x_1 (- $ is a dog$)(x_2)$
   c. $\forall x_2 \neg (- $ is a father$)(x_2) \vee \exists x_2 (- $ is related to $-)(x_2, $ Ralph$)$
   d. $\forall x_1 \forall x_2 ((- $ is a dog $-)(x_1) \vee (- $ is a cat$)(x_2)) \to (- $ barks$)($Juney$))$

Key Words predicate symbol   closed wff
               name symbol   open wff
               term   quantifier binds a variable
               metavariable   unique readability of wffs
               formal language   realization
               atomic wff   semi-formal language
               occurrence of a variable is free categorematic vocabulary
               atomic wff   logical vocabulary
               compound wff   universal closure of a wff
               scope of a quantifier   Form and Meaningfulness

10 A Predicate Applies to an Object or Objects

    A. Names 69
    B. Predications 70
    Exercises 74
    C. The Self-Reference Exclusion Principle 75
    D. The Universe for a Realization 76
    Exercises 77

Atomic propositions are primitive. They are structurally simple, consisting of names and what's left over when the names are deleted. They are also semantically simple: we agree to view each as true or false. So "(— is a dog) (Ralph)" is true or false. It's not for us as logicians to say which, or how to determine that. Our work is to give an analysis of complex propositions based on the semantic properties of the syntactic primitives.

A. Names

In English we use only a limited number of words as names, making it easier to recognize when a word is being used as a name. As it turns out, there's another person with the name "Richard L. Epstein" who is a linguist. This ambiguity precludes our using "Richard L. Epstein" in our reasoning because we can't treat it as a type that has always the same properties in a discussion. We should expect that if a word is used as a name, it names at most one object. But need it name any object? Consider:

    "Santa Claus is a dog"   false

    "Santa Claus is not a dog"   true

    "Santa Claus has a dog"   true? false?

Our intuition begins to falter when a name does not really pick out something.

    Whatever choice we make about how to reason with names like "Santa Claus" that don't refer except in a story, or names like "Homer" that might not pick out anything, will depend on how we reason with names that "really" pick out things. So let's assume that a name picks out some thing and only one thing. To simplify, we'll assume no further properties of names.

The Classical Abstraction of Names   If we use a word or phrase as a name, we assume that there is exactly one thing it picks out, and we take account of no other property of that word or phrase. We say a name *refers to* what it picks out and call that thing its *reference*.

B. Predications

We have the predicate "— is a dog". We can say it's true or false of the thing named by "Ralph" according to whether the atomic proposition formed with that predicate and name, "(— is a dog) (Ralph)", is true or false.

If we were to restrict logic to reasoning about only named objects, that is, objects for which we have names in the semi-formal language, we would not need to say anything more about the relation of predicates to things. But we want to reason about things that aren't named, such as all the pigs in Denmark, and about things it would be difficult to list names of, such as all people in the U.S. who had an appendectomy on May 19, 1990. We need to agree on what it means to say that "— is a dog" is true or false of an object whether named or unnamed.

Suppose I wish to communicate to you that the lamp on my desk weighs less than 2 kg. It has no name. If you were in the room with me I might point to the lamp and say, "That lamp weighs less than 2 kg." I use the words "That lamp" plus my gesture to pick out the thing, and hence by our standards, I use those words as a name. If I were writing to you, I might write "The brown lamp on my desk weighs less than 2 kg," using the phrase "The brown lamp on my desk" to pick out the one object. I name the object, however temporarily, and state a proposition using that name and the predicate "— weighs less than 2 kg".

If every time we wish to assert a connection between a predicate and an unnamed object we have to coin a new name and form a new semi-formal proposition not previously available to us, our semi-formal language will be always unfinished. We won't be able to get agreement in advance on the vocabulary we'll reason with.

Rather than accepting a proliferation of new names, we can use variables as temporary names: they await supplementation to pick out an object. For example, I could say "x weighs less than 2 kg" while I emphatically point to the lamp and utter x. Or I could write to you "x weighs less than 2 kg", and add "By 'x' I mean the lamp on my table". I still have to indicate what x is to refer to, by pointing or describing in words, but that way of indicating is separated from the semi-formal language.

When we stipulate by some means what x is to refer to and agree that "— weighs less than 2 kg" is a predicate, then "(— weighs less than 2 kg)(x)" is true or false. It is a proposition, indeed an atomic proposition.

The connection between atomic predicates and unnamed objects is reduced to naming and truth-values of atomic propositions.

We could just as well have used the variable y instead of x, in which case "(— weighs less than 2 kg)(y)" would provide "the" connection between the predicate and that object. Yet "(— weighs less than 2 kg)(y)" is a different formula and hence a different proposition from "(— weighs less than 2 kg)(y)", even though we agree to use both x and y to refer to the same object. Variables have form but no content except what we might assign to them via an indication of a particular reference. That method of assigning reference is necessarily outside our semi-formal language, connecting as it does a piece of language to "the world". Hence "(— weighs less than 2 kg) (*x*)" and "(— weighs less than 2 kg) (*y*)" should be *equivalent*: they have the same semantic properties, since we can't take account of their differences within the semi-formal language. One is true iff the other is.

But what if an object already has a name? Or several? For example, the person named "Marilyn Monroe" was also known by the names "Norma Jean Baker" and "Norma Jean Mortenson". If we also let "*x*" refer to that person, which of the following makes "the" connection between "— was an actress" and that object?

(— was an actress) (Marilyn Monroe)

(— was an actress) (Norma Jean Baker)

(— was an actress) (Norma Jean Mortenson)

(— was an actress) (*x*)

It shouldn't matter. We've assumed that the only property of a name is what it refers to, so the ways in which these names pick out the object can't be taken into account in our deliberations. It doesn't matter how we refer to a thing in making the connection between it and a predicate; all that matters is what the name, temporary or not, refers to. So in this context, each of these propositions is true or each is false.

What about *relations* (*n*-ary predicates for *n* > 1)? What shall we mean when we say that "— is bigger than —" is true of:

If we let "*x*" stand for the left-hand object and "*y*" for the right-hand one, then "(— is bigger than —) (*x*, *y*)" is true. If we let "*x*" stand for the right-hand object and "*y*" the left-hand one, then "(— is bigger than —) (*x*, *y*)" is false. It's not enough to speak of a relation applying to objects. We have to take account of which blank is to be filled with the name of which object. Except for that, the connection of relations to objects is the same as for unary predicates.

Predications Given an *n*-ary atomic predicate P and terms $t_1, \ldots, t_n$ with an indication of what each variable among those is to refer to, $P(t_1, \ldots, t_n)$ is a *predication*. We say that P is *predicated of* the objects referred to by the terms according to those references.

Example 1: (— loves —) (*x*, Ralph)
Analysis Suppose we let *x* stand for the lamp on my desk. Then, with the usual understanding of the words, this is false. If, however, we let *x* stand for me, the proposition is true.

Example 2: (— has the same blood type as —) (x, y)
Analysis If we let both x and y refer to the same object, say me, this is a true proposition. We could require that distinct variables stand for distinct things in a predication. But that would require an analysis of what we mean by two things being distinct, which in turn is greatly facilitated by allowing distinct variables to refer to the same object, as we'll see in Chapter 15. Besides, we allow different names to refer to the same object, like "Marilyn Monroe" and "Norma Jean Baker". So we'll allow distinct variables in a predication to refer to the same object.

Example 3: (— has the same blood type as —) (x, x)
Analysis Suppose we let the first x stand for me and the second stand for my dog Birta. Then . . . . We can't do that. Our agreement that a name can refer to only one thing applies to variables used as temporary names within a single predication.

Example 4: "(— was the teacher of —)" is true of Socrates and Plato.
Analysis Normally we label the blanks in a predicate in order from left to right, so to say that "— was the teacher of —" is true of Socrates and Plato, we understand the order of the objects to be that in which we mention them, first Socrates, then Plato. Then we can understand the example as meaning that "(— was the teacher of —) (x, y)" is true if we let x stand for Socrates and y stand for Plato.

> Here are other ways people say that a predicate is true of an object or objects:
>
> (— was the teacher of —) *applies to* Socrates and Plato (in that order)
>
> (— was the teacher of —) *is true of* Socrates and Plato (in that order)
>
> Socrates and Plato (in that order) *satisfy* (— was the teacher of —)
>
> (— was the teacher of —) *holds of* Socrates and Plato (in that order)
>
> Socrates *stands in relation of* (— was the teacher of —) *to* Plato

Thus, Socrates stands in the relation "— was the teacher of —" to Plato, though Plato does not stand in that relation to Socrates. When a predication is not true, we say it is *false of* or *is not satisfied* (etc.) of the objects in the order given in the predication.

Example 5: Stanisław Krajewski thinks that Marilyn Monroe was blonde.
   Stanisław Krajewski thinks that Norma Jean Baker was blonde.
Analysis When I asked Stanisław Krajewski, he said the first is true and the second is false. The propositions are not equivalent, though the only difference between them is that they use different names to name the same thing. The problem is that the truth or falsity of propositions that result from filling the blanks with names in "— thinks that — was blonde " depends on what names are used even if they refer to the same thing. Some semantic value of names, such as connotation, is significant here. But we've agreed that we won't take account of any other semantic values of names. So we have to exclude such predicates from our semi-formal languages.

Extensionality of predicates A predicate is *extensional* if given any terms $t_1, \ldots, t_n$ and $u_1, \ldots, u_n$ such that each term that is a variable is supplemented with an indication of what it is to refer to, and for each i, $1 \le i \le n$, t_i and u_i both refer to the same object, then $P(t_1, \ldots, t_n)$ has the same truth-value as $P(u_1, \ldots, u_n)$.

Only extensional predicates are allowed in a semi-formal language.

Example 6: If "— is a dog", and "Ralph", and "Marilyn Monroe", and "Norma Jean Baker" are in a realization with their usual meanings and references, then the following semi-formal wffs must all have the same truth-value.

(— is the dog of —) (x_1, x_2)	where x_1 refers to Ralph and x_2 to Marilyn Monroe
(— is the dog of —) $(x_1,$ Marilyn Monroe)	where x_1 refers to Ralph
(— is the dog of —) $(x_1,$ Norma Jean Baker)	where x_1 refers to Ralph
(— is the dog of —) (Ralph, x_2)	where x_2 refers to Marilyn Monroe
(— is the dog of —) (Ralph, Marilyn Monroe)	
(— is the dog of —) (Ralph, Norma Jean Baker)	
(— is the dog of —) (x_3, x_4)	where x_3 refers to Ralph and x_4 to Marilyn Monroe

In what follows, it is enough that we can agree that any atomic wff of a semi-formal language is (or represents what is) true or false when the free variables in it are given reference. How or why it is true or false is outside the scope of our studies. For that we look to the biologist, the economist, the mechanic, the nurse, the theologian. This is what we mean by saying that logic is independent of subject matter.

Aside: *The platonist conception of predicates and predication*
A platonist conceives of propositions as abstract objects, some of which can be "represented" or "expressed" in language. In accord with this view, the platonist says that predicates or properties or relations are abstract things, some of which can be represented by, or expressed as, or simply correspond to linguistic predicates as we've defined them. And a property or relation exists independently of our conceiving of it or representing it. Thus we have an inscription "is blue" that stands for a linguistic type (taken as an abstract thing) that corresponds (etc.) to the property of being blue, or "blueness". There may be properties and relations that do not correspond to any linguistic predicate in our language.

Different linguistic expressions may represent the same abstract predicate, for example, "(— is a bachelor)" and "(— is a man) ∧ ¬(— is married)" where both blanks in the latter are meant to be used to apply to the same object. Whereas we would undertake an analysis of the linguistic expressions to decide whether we should treat these as equivalent, the platonist goes through much the same procedure to determine whether the expressions "denote" the same predicate.

"A unary predicate applies to an object" is understood by a platonist to mean that (or simply is true just in case) the object has the property. For example, the predicate denoted by "— is a dog" applies to Ralph just in case Ralph has the property of being a dog; the predicate expressed by "— is a horse" is true of Dusty just in case the quality of horsieness belongs to Dusty. How we indicate what a variable is to refer to is of no concern, nor what name we use, for an object has the property or it does not, and it is an accident how or whether we refer to the object with a variable or name.

Nor, the platonist says, need we concern ourselves with what variable is to refer to what object in the predication of a relation. The "concreta" Socrates and Plato "participate in" the relation expressed by "— was the teacher of —" in the order represented by the sequence (Socrates, Plato). A sequence of objects is an abstract object, not our way of assigning reference. Sometimes it is said that the sequence (Socrates, Plato) satisfies the predicate represented by "— was the teacher of —", and that is if and only if Socrates stands in the relation of being the teacher of to Plato.

Thus, a platonist talks of doggieness or blueness or the relation of being the father of, where we speak of the linguistic predicates "— is a dog", "— is blue", "— is the father of —". The platonist explains these odd locutions as a deficiency, indeed a necessary deficiency of language in discussing abstract things.

Propositions are no longer irreducible, but what they are composed of must be abstract. Properties and relations are abstract things. A sequence, even of just one object, is an abstract thing. A proposition is "composed" of a property or relation and a sequence. What exactly that composition is seems unclear, since it can't be juxtaposition, for abstract things aren't like concrete things that can be placed side by side. Nor can it be explained by saying that the composition is the objects having the property because that doesn't account for false propositions.

Others take a more psychological approach and say that a proposition is composed of concepts of individuals and concepts of predicates. The composition is itself a concept or thought, a thought which each of us can have but which exists independently of us all, and hence is in some sense abstract.

The platonist view of predicates is recommended by its adherents for giving an explanation of truth and objectivity that is independent of us and our capabilities. But it does so at the cost of separating logic from reasoning. To use logic we shall still have to indicate what variable is to refer to what thing and how we shall specify sequences.

Nonetheless, the logic we develop here can be adopted by the platonist by treating what we call predicates as representatives of "real" predicates. The explanation the platonist gives of what makes an atomic proposition true is just one of many that is compatible with our work here.

Exercises
1. What semantic properties do we assume a name has?
2. a. State precisely what it means to predicate the phrase "— is round" of: △
   b. State precisely what it means for the predicate "— stands to the left of —" to apply to: △ ◯

c. State precisely what it means to say that the predicate "— and — are larger than —" is true of: △ ○ □

d. State precisely what it means to say that Marilyn Monroe stands in the relation "— is more honest than —" to Donald Trump.

3. a. What does it mean to say that a predicate is extensional?
   b. Give two examples of non-extensional predicates not mentioned in the text.
   c. Why do we require that every atomic predicate in a semi-formal language be extensional?

4. a. Are the following necessarily equivalent?
      i. " (— is a dog) (x)" where "x" is meant to stand for Ralph
      ii. " (— is a dog) (y)" where "y" is meant to stand for Ralph
      iii. Ralph is a dog
   b. What role do (i)–(iii) play in predicating "—is a dog" of Ralph?
   c. Instead of (i) why don't I write: "" (— is a dog) (x)" where "x" is meant to stand for "Ralph" " ?

C. Self-Reference

If Dick says, "I'm over 6 feet tall" he's talking about himself. If Zoe says, "That's my cup of coffee", she is talking about herself. Self-reference is an apparently essential part of our language that reflects our self-consciousness: without "I" and "my" we can be in the world but not express our knowledge of that. Yet the power of self-reference within our language can create puzzling problems.

Epimenides the Cretan is reported to have said, "All Cretans are liars." Was he speaking truly? This is called *the liar paradox*.

Suppose we name each wff of the formal language (it's not hard to come up with a method). Suppose that the following sentence is called "Urx":

(— is false) (Urx)

If Urx is true, then since it says that it's false, it's false. So Urx has to be false. But then since that's what it says, Urx has to be true. How can we then give truth values to the atomic predications?

People have been worried about versions of the liar paradox for more than 2,000 years. There's no easy solution. We'll sidestep the problem by making the following assumption.

The Self-Reference Exclusion Principle No name symbol can be realized as a name of any wff or part of a wff of the semi-formal language. No predicate symbol can be realized as a predicate that can be used to define the truth or falsity of any wff (with references for variables supplied) of the semi-formal language.

Excluding self-referential sentences from realizations enforces a sharp distinction between logic and *metalogic*, between propositions we shall reason with and propositions about those propositions

D. The Universe for a Realization

In choosing the predicates that realize the predicate symbols in a realization, we have some idea of what objects we are discussing, since we agree that whenever the blanks in a predicate are filled with names of things we are talking about, the resulting sentence becomes a proposition. We should be explicit and state as precisely as we can what things are under discussion. Those things comprise the *universe* for the realization.

When we choose a collection of things to be the universe for a realization, we must have some idea of what it means to pick out something from it to be the reference of a variable. We might say that for a universe of pigs and dogs, pointing will do. Though we can't point to each pig in Denmark, we understand what it would mean to pick out one by pointing, and that's what counts as giving reference to a variable.

If someone proposes that we take the universe to be the counting numbers understood as abstract objects, she might suggest that we can point to them via their names: 1, 2, 3, ...

If someone suggests that we take all real numbers to be the universe for a realization, understood as abstract things, we can't point to them physically or with names, for there are—as shown by a mathematical argument—more real numbers than we could name even in theory. The platonist might say that we point to real numbers with our intellect and not our fingers: we mentally attach labels. Others might say that we can name real numbers with constructive decimal expansions. Or someone might say we can point to real numbers by equations that pick them out as solutions.

A physicist might wish to take as universe all the atoms in this piece of paper. We can't (apparently) point to those. Can we conceptually? But atoms are supposed to be physical. To the extent that we can "name" atoms via co-ordinate systems and numerical representations, perhaps we can point.

If we propose a collection of things as the universe for a realization, we are under an obligation to say what it means to pick out an object as a reference of a variable. Unless we can do that, all that follows here will be an empty formalism, unconnected with truth and reasoning.

Why don't we just choose the universe to be all things? After all, logic is supposed to be neutral with respect to what things there are. If we were to do so, we'd have to have some very general explanation of what it means to distinguish one thing from all others, assigning it as reference to a variable. To give such an explanation amounts to coming up with a univocal idea of what it means to be a thing. But that is what we hope to clarify by our development of predicate logic.

Chapter 10 Semantics for Classical Predicate Logic

But worse, as you can see in Appendix 3, the notion of all things is incoherent. So we'll work with "restricted" universes.

As an example, consider the realization of Example 5 of Chapter 9:

$$L(\neg, \rightarrow, \wedge, \vee, \forall, \exists, P_0, P_1, \ldots, c_0, c_1, \ldots)$$
$$\downarrow$$
$$L(\neg, \rightarrow, \wedge, \vee, \forall, \exists\,; \; -\text{ is a dog}, \; -\text{ is a cat}, \; -\text{ eats grass},$$
$$-\text{ is a wombat}, \; -\text{ is the father of } -\,; \text{ Ralph, Dusty, Howie, Juney })$$

Here are some choices for what we could take as the universe for this realization:

a. All animals alive in 1986 or toy animals.
b. All animals alive in 1986 in the U.S. that are not in a zoo or are toy animals.
c. All material things in the U.S. in 1986 over 3 cm tall.
d. The collection composed of just Ralph, Dusty, Howie, and Juney.

With universe (a), "$\exists x\, (x$ is a wombat)" is true, while with universe (b), that wff is (I suspect) false. With different universes, different wffs are true.

If we have names in the semi-formal language, then by the assumption that names refer there must be at least one object in the universe. If only predicate symbols are realized, however, it seems there needn't be anything in the universe. But if there's nothing in the universe, then there's nothing to take as reference of a variable, and our whole story of naming and predication becomes nonsense. We shouldn't investigate whether there are any wild zebras in the U.S. by taking the collection of all wild zebras in the U.S. as the universe for a realization, for we can embed that question in discussions within the semi-formal language by using, for example, "$\exists x\, [\, (-\text{ is wild})\, (x) \wedge (-\text{ is a zebra})\, (x) \wedge (-\text{ is in the U.S.})\, (x))\,]$" with universe all animals in the U.S. Explicitly, we adopt the following.

Nonempty Universe There is at least one thing in the universe for a realization.

Since the things that variables can stand for are just the objects in the universe, we say that the universe comprises the *range of the variables*.

Exercises

1. Show that the liar paradox as supposedly said by Epimenides the Cretan is not a paradox as stated. How can we make it paradoxical?
2. a. Why do we adopt the Self-Reference Exclusion Principle?
   b. Give two examples of sentences we would like to reason with but which are excluded from semi-formal languages by the Self-Reference Exclusion Principle.
3. Why don't we accept the collection of all things as a universe for a realization?

78 *An Introduction to Formal Logic*

4. Why do we require the universe for a realization to contain at least one thing?
5. Given an example where the truth-value of a semi-formal wff can depend on what universe is chosen for the realization.
6. We wish to use logic in physics. Give a story of what it means to assign a particular atom as reference to a variable. Does your story have more in common with assigning reference when discussing abstract objects such as real numbers or when discussing physical objects such as chairs and tables? Are atoms physical?

Key Words   The Classical Abstraction of Names
reference
predication
equivalent propositions
extensional predicate
self-reference
The Self-Reference Exclusion Principle
universe for a realization
nonempty universe
range of the variables

11 Models for Classical Predicate Logic

A. Assignments of References 79
B. Valuations 81
C. Compositionality and the Division of Form and Content . . 81
D. The Truth-Value of a Complex Proposition 82
E. Models 85
F. Tautologies and Semantic Consequence 87
Exercises 89
G. Propositional Logic in Predicate Logic 89
H. The Division of Form and Content Verified 91

A. Assignments of References

Suppose we have a realization and a universe of objects for it. We've already agreed on some relation between those.

- For each name, a specific object of the universe is taken as reference. How that object is picked out is not for us as logicians to say.

- The method whereby we indicate which object of the universe is to be used as (temporary) reference of which variable is not for us as logicians to specify it, but only to assume that such a method is given, as indeed we already have in selecting a universe for the realization. The method itself does not give any particular references for variables. It only specifies how such references can be provided.

Let's use the *metavariables* σ (sigma), $\sigma_0, \sigma_1, \ldots, \tau$ (tau), $\tau_0, \tau_1, \ldots$, and γ (gamma) to stand for ways of assigning references to both the variables and the names in accord with the method. Since the names are to be used as having fixed reference, each of these particular ways should give the same reference for a name. That is, if c is a name in the realization, then for any σ and τ, $\sigma(c) = \tau(c)$.

Example 1: Here is a realization and universe we saw before:
L($\neg, \rightarrow, \wedge, \vee, \forall, \exists$; — is a dog, — is a cat, — eats grass,
— is a wombat, — is the father of —; Ralph, Dusty, Howie, Juney)
universe: all animals, living or toy

Here are two assignments of references we could have with it.

σ("Ralph") = Ralph
σ("Dusty") = Dusty
σ("Howie") = Howie
σ("Juney") = Juney
$\sigma(x_0)$ = my dog Bidu
$\sigma(x_1)$ = the black dog that lives at my neighbor's across the road

$\sigma(x_2)$ = the brown cow in the field next to my home
$\sigma(x_3)$ = the white bull in the field next to my home
$\sigma(x_4)$ = the white calf in the field next to my home
$\sigma(x_5)$ = Juney
$\sigma(x_6)$ = the black dog that lives across the road
*   *   *   *   *   *   *   *
$\tau(\text{``Ralph''})$ = Ralph
$\tau(\text{``Dusty''})$ = Dusty
$\tau(\text{``Howie''})$ = Howie
$\tau(\text{``Juney''})$ = Juney
$\tau(x_0)$ = the cat my landlord has
$\tau(x_1)$ = the white calf in the field next to my home
$\tau(x_2)$ = my toy wind-up sheep
$\tau(x_3)$ = the black dog that lives at my neighbor's across the road
$\tau(x_4)$ = the white calf in the field next to my home
$\tau(x_5)$ = the white bull in the field next to my home
$\tau(x_6)$ = the brown cow in the field next to my home
$\tau(x_{811})$ = the biggest elephant in the Albuquerque zoo
$\tau(x_{916})$ = the smallest wolf in the Albuquerque zoo
$\tau(x_{4318})$ = the beaver that lives in the canal west of my home

Analysis Two assignments can assign the same object to a variable: $\sigma(x_4) = \tau(x_4)$, or they can assign different ones, $\sigma(x_2) \neq \tau(x_2)$. An assignment can assign the same object to distinct variables: $\sigma(x_1) = \sigma(x_6)$. And an assignment can assign to a variable an object that has a name : $\sigma(x_5)$ = Juney. We need to allow that so we can say that "$\exists x_5 ((-\text{ is a dog}) (x_5))$" is true by saying "let x_5 stand for Juney".

For any one predication or proposition, only the variables appearing in it will be of concern to us. But it is convenient to imagine that each variable is or can be assigned an object by each assignment of references, for we can always extend an assignment as needed.

If our variables are to express generality, the method of assigning reference has to allow us to pick any object of the universe as the reference of any variable. We don't want "$\exists x_3 ((-\text{ is a wombat}) (x_3))$" to come out false because no assignment of references assigns a wombat to x_3.

Completeness of the collection of assignments of references There is at least one assignment of references. For every assignment of references σ, and every variable x, and every object of the universe, either σ assigns that object to x or there is an assignment τ that differs from σ only in that it assigns that object to x.

We write $\tau \sim_x \sigma$ to mean that the assignment of references τ differs from σ at most in what it assigns to x.

Chapter 11 Models for Classical Predicate Logic 81

We've assumed that there is at least one object in the universe, so there's at least one assignment of references.

B. Valuations of atomic wffs
We describe a way the world could be, according to our assumptions, with a realization, a universe, a complete collection of assignments of references, and a specification of truth-values to all the atomic propositions, which include the open atomic wffs when the variables in them are assigned reference.

Valuations Given an n-ary predicate P and terms $t_1, \ldots, t_n$ and an assignment of references σ, a truth-value is given for $P(t_1, \ldots, t_n)$. The evaluations of truth-values for all atomic predications relative to σ is called the *valuation* $\vee_\sigma$ *on atomic wffs based on* σ. We write:
$$\vee_\sigma(P(t_1, \ldots, t_n)) = \mathsf{T} \text{ or } \vee_\sigma(P(t_1, \ldots, t_n)) = \mathsf{F}.$$
Sometimes we write instead: $\vee_\sigma \vDash P(t_1, \ldots, t_n)$ or $\vee_\sigma \nvDash P(t_1, \ldots, t_n)$.

Example 2: (— is a wombat) (Ralph)
Analysis For the realization and universe of Example 1 with a complete collection of assignments, "Ralph" is assigned Ralph. Since Ralph isn't a wombat, let's assign this wff F. This doesn't depend on any particular assignment of references since for every γ, we have $\gamma(\text{"Ralph"}) = \text{Ralph}$. So for every γ, $\vee_\gamma(\,(\text{— is a wombat})\,(\text{Ralph})\,) = \mathsf{F}$. But more, for every x, if $\gamma(x) = \text{Ralph}$, then $\vee_\gamma(\,(\text{— is a wombat})\,(x)\,) = \mathsf{F}$, since our predicates are extensional: it doesn't matter how we name an object.

We could say instead that the example is true, which would be like saying, "Suppose Ralph were a wombat". That would be a different way the world could be. We could say that "Ralph" refers to my puppy Arfito, so that for each assignment γ, $\gamma(\text{"Ralph"}) = \text{Arfito}$, which would be like saying "Suppose Ralph were Arfito". That, too, would be a different way the world could be.

There will be as many valuations as there are ways to assign references. Together they must accord with the assumption we made that the predicates in the semi-formal language are understood as extensional.

Extensionality of Atomic Predications For any atomic wffs $P(t_1, \ldots, t_n)$ and $Q(u_1, \ldots, u_n)$ and assignments σ and τ, if for all i, $\sigma(t_i) = \tau(u_i)$, then
$$\vee_\sigma(P(t_1, \ldots, t_n)) = \vee_\tau(Q(u_1, \ldots, u_n)).$$

Letting semi-formal wffs name themselves, I'll write, for example, $\vee_\sigma((\text{—is a dog})(\text{Ralph}))$ rather than $\vee_\sigma(\text{"(—is a dog)(Ralph)"})$.

C. Compositionality and the Division of Form and Content
The compositionality assumption in propositional logic says that the truth-value of a proposition is determined by its form, as given by the connectives appearing

in it, and the semantic properties of its constituents. For predicate logic shall we say that the truth-value of the whole is determined by the form of the proposition, as given by the logical vocabulary appearing in it, and the properties of its categorematic parts?

This fails to grasp the structure of propositions. Predicates have form, unlike atomic propositions in propositional logic: a binary predicate is different from a ternary one. So the form of a proposition depends on the predicates appearing in it, too. But even adjusting for that difference, the proposal is not right. The truth-value of "$\forall x_5 ((-\text{ is a dog})(x_5))$" depends not only on the predicate "— is a dog", the quantifier, and the variable, but on what things there are. Is there anything further on which the truth-value of a complex proposition could depend? Any other factor would be extraneous to what we have explicitly chosen to take into account.

Compositionality The truth-value of a complex proposition is determined by its form, the semantic properties of its parts, and what things there are.

In propositional logic we made a sharp distinction between the role that form plays and the semantic properties of propositions and their constituents. With the same motivation, we adopt the following, though here we have to speak not only of propositions but of any part of a proposition.

The Division of Form and Content If two propositions or parts of propositions have the same semantic properties, then they are indistinguishable in any semantic analysis, regardless of their form.

D. The Truth-Value of a Complex Proposition

Our understanding of "all" and "some" will be the basis for how we interpret $\forall$ and $\exists$.

Example 3: *All dogs bark.*
Analysis This is true iff every single dog barks, no exceptions.

Example 4: *All polar bears in Antarctica can swim.*
Analysis There are no exceptions: there's not one polar bear in Antarctica that can't swim. There also aren't any polar bears in Antarctica that can swim. There aren't any polar bears at all in Antarctica. So is the example true?

Some people say the claim is false: there has to be at least one thing for us to be right when we say "all" in ordinary conversation. Others say the claim is true, since there are no exceptions. We have a choice.

Interpreting $\forall$ Our formalization of "all" as $\forall$ is to be understood as meaning each and every one, no exceptions, which includes the possibility that there might not be any.

Example 5: Some dog barks.
Analysis This is true: my dog Birta barks.

Example 6: Dr. E: *At the end of this term, some of my students will get an A.*
Analysis At the end of the term one student in all of Dr. E's classes got an A . Was Dr. E right? If you don't think so, then how many is "some students"? At least 2? At least 8? At least 10%? More than 18%? When we say "some," we're only guaranteeing that there is at least one.

Example 7: Dr. E: *Some of my students will pass my next exam.*
Analysis All Dr. E's students pass the exam. Was Dr. E right? For this claim to be true, don't some students also have to fail? With "some" we often mean "at least one, but not all". We have a choice.

Interpreting $\exists$ Our formalization of "some" as $\exists$ is to be understood is to be understood as meaning at least one and possibly all.

Later we'll see how we can formalize within our system other readings of "some".

To see how we can use these readings of $\forall$ and $\exists$ in our formal system, let's assume we have a realization and universe along with assignments of references and for each of those a valuation for the atomic predications. We want to extend those valuations to an evaluation v for all closed wffs. I'll let "x" stand for "x_1" and "y" stand for "x_2" in the examples.

Example 8: "$\forall x ((-\text{ is a dog})(x)$" is true
 iff (a) (informally) everything is a dog
 iff (b) no matter what thing in the universe x refers to, "$(-\text{ is a dog})(x)$" is true
 iff (c) for any assignment of references σ, $v_\sigma((-\text{ is a dog})(x)) = T$

Example 9: "$\exists x (-\text{ is a dog})(x)$" is true
 iff (a) (informally) something is a dog
 iff (b) there is something in the universe such that if x is given an indication that it is to refer to that thing, "$(-\text{ is a dog})(x)$" is true
 iff (c) there is some assignment of references σ such that $v_\sigma((-\text{ is a dog})(x)) = T$

These evaluations emphasize that it is not the existence of some object or all objects that satisfy a predicate we consider in determining the truth-value of the wff, but that the object(s) are picked out. We replace "for all objects" with "for all ways to pick out an object"; we replace "there exists an object" with "there is some way to pick out an object". Our analysis of variables and quantification would be just an empty formalism if we didn't take into account pointing and naming, however "theoretical" those may be.

Example 10: $(-\text{ is a dog})(\text{Ralph})$
Analysis The evaluation of this will be the same regardless of which assignment of

references we use, since all give the same assignment to "Ralph". Hence,
"(— is a dog) (Ralph)" is true iff for every assignment of references σ,
$$v_\sigma(\,(\,—\text{ is a dog})\,(\text{Ralph})\,) = T$$
iff for some assignment of references σ,
$$v_\sigma(\,(\,—\text{ is a dog})\,(\text{Ralph})\,) = T$$

Example 11: $v_\sigma(\,(\,—\text{ is a dog})\,(x) \to \neg\,(\,—\text{ is a cat})\,(x)\,) = T$
 iff $v_\sigma(\,(\,—\text{ is a dog})\,(x) \to \neg\,(\,—\text{ is a cat})\,(x)\,) = T$
 iff $v_\sigma(\,(\,—\text{ is a dog})\,(x)\,) = F$ or $v_\sigma(\neg\,(\,—\text{ is a cat})\,(x)\,) = T$
 iff $v_\sigma(\,(\,—\text{ is a dog})\,(x)\,] = F$ or $v_\sigma(\,(\,—\text{ is a cat})\,(x)\,) = F$

Analysis We are justified in evaluating the propositional connectives as we did in classical propositional logic because relative to an assignment of references σ, an open formula such as "(— is a dog) (x)" is a proposition, true or false.

Assignments of references provide the link that justifies the use of propositional connectives between open wffs.

Example 12: "$\forall x\,\exists y\,(\,—\text{ is the father of }—)\,(y, x)$" is true
 iff no matter what thing x refers to,
  "$\exists y\,(\,—\text{ is the father of }—)\,(y, x)$" is true
 iff no matter what thing x refers to, there is some thing such that if the use of y is supplemented with an indication that y is to refer to that thing,
  "(— is the father of —) (y, x)" is true

We have a problem stating this in terms of assignments of references. If we let σ assign reference to x, then we've also given an assignment to y, since we've assumed that each assignment gives reference to every variable. So we let σ assign a reference to x. Then we're free to choose another assignment of references τ that agrees with σ for all variables except possibly y. We can do that because the collection of assignments of references is complete. So we have:

"$\forall x\,\exists y\,(\,—\text{ is the father of }—)\,(y, x)$" is true
 iff for any assignment of references σ,
  $v_\sigma(\,\exists y\,(\,—\text{ is the father of }—)\,(y, x)\,) = T$
 iff for any assignment of references σ, there is some assignment of
  references τ such that $\tau \sim_x \sigma$ and $v_\tau(\,(\,—\text{ is the father of }—)\,(y, x)\,) = T$

In this way *at each stage in an analysis of the truth-value of a complex proposition we have to account for assigning reference to at most a finite number of occurrences of a single variable not assigned reference before.*

Example 13: "$\forall x\,\forall y\,(\,—\text{ is a cousin of }—)\,(x, y)$" is true
 iff for any assignment of references σ,
  $v_\sigma(\,\forall y\,(\,—\text{ is a cousin of }—)\,(x, y)\,) = T$
 iff for any assignment of references σ, and any assignment of
  references τ such that $\tau \sim_y \sigma$, $v_\tau(\,(\,—\text{ is a cousin of }—)\,(x, y)\,) = T$
 iff for any assignment of references γ, $v_\gamma(\,(\,—\text{ is a cousin of }—)\,(x, y)\,) = T$

Analysis The last equivalence follows from the completeness of the collection of assignments of references, since given any two objects there is some assignment of references that assign one to x and the other to y.

Example 14: "$\forall x\,((- \text{ is a dog})\,(x) \vee \exists x\,(- \text{ eats grass})\,(x))$" is true
  iff for any assignment of references σ,
  $\quad v_\sigma((- \text{ is a dog})\,(x) \vee \exists x\,(- \text{ eats grass})\,(x)) = T$
  iff for any assignment of references σ,
  $\quad v_\sigma((- \text{ is a dog})\,(x)) = T$ or $v_\sigma(\exists x\,(- \text{ eats grass})\,(x)) = T$
  iff for any assignment of references σ, $v_\sigma((- \text{ is a dog})\,(x)) = T$ or
  there is some τ such that $v_\tau((- \text{ eats grass})\,(x)) = T$

Analysis To evaluate the first quantifier $\forall x$, we use σ. But that cannot be used to give a reference for x in "$(- \text{ eats grass})\,(x)$" because that x is not bound by $\forall x$ but by the quantifier $\exists x$, for which we use the assignment τ.

In the formal language we build up a formula in steps. In the evaluation of a formula we unpack the formula in the reverse order until we reach the lowest level: atomic wffs. An assignment of references allows us to treat those as propositions. Let's set out this procedure generally.

E. Models

We begin with a formal language. We realize (some of) the predicate symbols and name symbols and have a semi-formal language. We agree on a universe for the semi-formal language. We provide an *interpretation* of the semi-formal language: assignments of references and valuations for atomic wffs. So for every atomic wff of the semi-formal language $P(t_1, \ldots, t_n)$ and for every assignment of references σ, $v_\sigma(P(t_1, \ldots, t_n))$ is defined. We now extend simultaneously all the valuations v_σ to all formulas of the semi-formal language.

The extension of all valuations to all wffs
For all σ, for every wff A of the semi-formal language:

$\quad v_\sigma(\neg A) = T \quad$ iff $\quad v_\sigma(A) = F$

$\quad v_\sigma(A \wedge B) = T$ iff $v_\sigma(A) = T$ and $v_\sigma(B) = T$

$\quad v_\sigma(A \vee B) = T$ iff $v_\sigma(A) = T$ or $v_\sigma(B) = T$

$\quad v_\sigma(A \to B) = T$ iff $v_\sigma(A) = F$ or $v_\sigma(B) = T$

$\quad v_\sigma(\exists x\,A) = T \quad$ iff for some assignment of references τ such that
$\quad\quad\quad\quad\quad\quad\quad\quad \tau \sim_x \sigma,\ v_\tau(A) = T \quad\quad$ *the classical evaluation of $\exists$*

$\quad v_\sigma(\forall x\,A) = T \quad$ iff for every assignment of references τ such that
$\quad\quad\quad\quad\quad\quad\quad\quad \tau \sim_x \sigma,\ v_\tau(A) = T \quad\quad$ *the classical evaluation of $\forall$*

If $v_\sigma(A) = T$, we say that σ *satisfies* or *validates* A, and we write $v_\sigma \vDash A$ or $\sigma \vDash A$. If $v_\sigma(A) = F$, we write $v_\sigma \nvDash A$ or $\sigma \nvDash A$.

Theorem 1 For every wff A and for every assignment of references σ, $\vee_\sigma(A)$ is defined.

Proof We proceed by induction on the length of wffs (see Appendix 1 for an explanation of proof by induction). For $n = 1$, A is an atomic wff, and for each assignment of references σ, $\vee_\sigma(A)$ is defined by the interpretation.

Now suppose A has length $n + 1$. Then by the unique readability of wffs, A has one and only one of the following forms: $\neg B$, $B \to C$, $B \wedge C$, $B \vee C$, $\forall x\, B$, or $\exists x\, B$, where B and C are unique.

If A is of the form $\neg B$, then B has length $\leq n$, so by induction, $\vee_\sigma(B)$ is defined. So by our method of evaluation, $\vee_\sigma(A)$ is defined. The cases when A has the form $B \to C$, or $B \wedge C$, or $B \vee C$ follow in the same way.

Now suppose that A has the form $\forall x\, B$. Then $\vee_\sigma(\forall x\, B) = T$ iff for every assignment of references τ such that $\tau \sim_x \sigma$, $\vee_\tau(B) = T$. Since B has length $\leq n$, by induction $\vee_\tau(B) = T$ is defined for all τ. Hence, $\vee_\sigma(\forall x\, B)$ is defined.

The case when A has form $\exists x\, B$ is done similarly. ∎

In a model we have $\vee_\sigma(A)$ defined for every σ and every formula A. The formulas that are propositions, however, are closed ones.

The valuation $\vee$ on all closed wffs For every closed wff A of the semi-formal language, $\vee(A) = T \equiv_{Def}$ for every assignment of references σ, $\vee_\sigma(A) = T$.

A model A *model* is a realization, a universe for the realization, a complete collection of assignments of references, valuations for atomic wffs satisfying the consistency conditions, the extension of the valuations to all wffs by the inductive definition, and the valuation on all closed wffs. A proposition A of the semi-formal language is *true in the model* iff $\vee(A) = T$. Otherwise it is *false in the model*, and we write $\vee(A) = F$.

I'll use $M, M_0, M_1, \ldots$ to stand for models. Sometimes we write $M \vDash A$ for $\vee(A) = T$, and $M \nvDash A$ for $\vee(A) = F$. Schematically, a model is:

$$L(\neg, \to, \wedge, \vee, \forall, \exists, P_0, P_1, \ldots\ldots, c_0, c_1, \ldots)$$
$$\downarrow \text{ realization}$$
$L(\neg, \to, \wedge, \vee, \forall, \exists$, realizations of predicate symbols and name symbols)
universe: specified in some manner

$\downarrow$ assignments of references, $\sigma, \tau, \ldots$; assignments of truth-values to atomic predications, $\vee_\sigma, \vee_\tau, \ldots$;
$\{T, F\}$ truth-tables for $\neg, \to, \wedge, \vee$; evaluation of the quantifiers

Aside: Pure reference and truth
To use a model together we must:
     Agree on what we mean by naming and distinguishing things in the universe.
     Agree on the truth-values we assign to the atomic propositions.

For our logic as a whole, we need only some general conception of truth and naming and distinguishing to give us the structural constraints on our language and models.

Platonists say even less is needed. They hold that there is some undefined and perhaps undefinable way in which predicates are true of objects independent of us. For them, simply:

"$\forall x$ (— is a dog) (x)" is true of this universe iff

the predicate represented by "(— is a dog)" is true of every object in the universe

They view talk of assignments of references as just a way to say this more tractably for an inductive definition of truth for complex formulas. Nothing about distinguishability or picking out things need be said. Each thing in the universe is distinct from all others, but that is not connected to us, only to some abstract or theoretical notion of distinguishing, perhaps in Plato's heaven or by the hand of God. Simply, there are ways to assign reference, whether we recognize or consider them.

I see that view as an abstraction from how we talk and reason. They say that I muddle the pure reality of reference and truth. But if the pure existence of such assignments of references and truth is not tied to our methods of naming-pointing-distinguishing, then I cannot understand what those assignments are and I have no idea how we could use the formal systems to evaluate reasoning.

F. Tautologies and Semantic Consequence

We adopt the same definitions of tautology and semantic consequence as for classical propositional logic, relative to the language and models here.

Tautologies A wff is a *tautology* or *valid* iff in every model its realization is evaluated as true. In that case we write $\models A$.

A semi-formal proposition is a tautology iff it is the realization of a wff that is a tautology.

A proposition in ordinary English is a tautology if there is a good formalization of it that is a tautology.

A scheme of wffs is a tautology iff every wff of that form is a tautology.

Semantic consequence For a collection of formal wffs Γ and formal wff A, the inference Γ *therefore* A is *valid* means that there is no model in which all the wffs in Γ are true and A is false. In that case we say that A is a *semantic consequence* of Γ, or that the pair Γ, A is a *semantic consequence*, and write $\Gamma \models A$.

A semi-formal inference is valid iff it is the realization of a formal inference that is valid. An inference in ordinary English is valid if there is a good formalization of it which is valid.

If M is a model in which all the wffs of Γ are true, we say M is a *model of* Γ and write $M \models \Gamma$.

Analysis This has the form A ∨ ¬ A, where A is a closed wff. Since we evaluate the propositional connectives as we did in classical propositional logic, this will be true in every model. It's a tautology.

Example 15: ∃x ((— is a dog) (x) ∨ ¬ (— is a dog) (x))
Analysis This, too, is a tautology because the scope of the quantifier has the form A ∨ ¬ A—though here A is an open wff. In any model relative to any assignment of references, the scope of the quantifier will be evaluated as true.

Example 16: ∃x (— is a dog) (x) → ¬ (— is a cat) (Ralph)
Analysis This is not a tautology: we could have a model in which $\sigma(x)$ is Fido, a dog, and so ⋁ ((— is a dog) (x)) = T, and say that "Ralph" denotes some particular cat, say Howie, so that ⋁ ((— is a cat) (Ralph)) = T.

Example 17: ∀x (— is a dog) (x) → (— is a dog) (Ralph)
Analysis This is a tautology. If the antecedent is true in a model, then for any assignment of references σ, $⋁_\sigma$ ((— is a dog) (x)) = T. So in particular, for an assignment of references τ that assigns to x the same object it assigns to "Ralph", $⋁_\tau$ ((— is a dog) (x)) = T. So $⋁_\tau$ ((— is a dog) (Ralph)) = T.

Example 18: ∀x (— is a dog) (x)
   Therefore, (— is a dog) (Ralph)
Analysis This is a valid inference. If a wff of the form ∀x A is true in a model, then for every name c in the realization, A with c replacing x is true—after all, the wff is true for every object, so it's got to be true for the object named by c.

We can also justify this as a tautology because the last example is a tautology and the Semantic Deduction Theorem holds here: A→ B is a tautology iff A⊨ B, where A and B are closed wffs.

The properties of the semantic consequence relation in classical propositional logic that we saw in Theorem 1, p. 31, hold for classical predicate logic. Those do not depend on the particular language or models that are employed.

Example 19: ∀x ((— is a dog) (x) → (— barks) (x)
   (— is a dog) (Ralph)
   Therefore, (— barks) (Ralph)
Analysis This is a valid inference.

Example 20: (— is a dog) (Ralph)
   Therefore, ∃x (— is a dog) (x)
Analysis This is a valid inference: every σ assigns the same object to "Ralph"; for some assignment τ, $\tau(x)$ = the object assigned to "Ralph"; and $⋁_\tau$ ((— is a dog) (Ralph)) = T. So it follows that $⋁_\tau$ ((— is a dog) (x)) = T.

We've reduced the notion of a way the world could be to that of a model. For this to provide a good formalization of the notion of validity we need to have enough models to allow for any possibility relative to what we're paying attention to in our syntax and semantics. Then we can define our logic.

Sufficiency of the collection of models For any realization and any universe and complete collection of assignments of references, any assignment of truth-values to the atomic predications satisfying the extensionality condition defines a model.

Classical predicate logic *Classical predicate logic* comprises the formal language along with the definitions of realization, models, tautology, and semantic consequence.

Exercises (In the following, x stands for x_1 and y stands for x_2.)
1. Write out the evaluation of the following to show that it is valid:
   $\exists x\,((-\text{ is a dog})(x) \vee \neg(-\text{ is a dog})(x))$
2. Assuming a model, give the evaluation of the following propositions.
   a. $\forall x\,((-\text{ loves Juney})(x) \vee \neg((-\text{has a heart})(x)))$
   b. $\exists x\,((-\text{ is a cat})(x) \vee (-\text{ is a dog})(\text{Ralph}))$
   c. $\forall x\,((-\text{ is a dog})(x) \rightarrow \neg((-\text{ eats grass})(x)))$
   d. $\forall x\,\exists y\,((-\text{ is the father of }-)(x,y))$
   e. $\exists y\,\forall x\,((-\text{ is the father of }-)(x,y))$
   f. $\exists x\,\forall y\,((-\text{ is the father of }-)(x,y))$
   g. $\forall x\,\forall y\,((-\text{ is the father of }-)(x,y) \vee \exists y\,(-\text{is a clone})(y))$
   h. $\forall x\,\neg(\exists y\,(-\text{ loves }-)(x,y) \wedge \neg\exists y\,(-\text{ loves }-)(x,y))$
3. Show directly each of the following for every model in which these (abbreviated) wffs appear.
   a. $\upsilon(\exists x\,\exists y\,((-\text{ is taller than }-)(x,y))) = T$ iff
      for some γ, $\upsilon_\gamma((-\text{ is taller than }-)(x,y)) = T$
   a. $\upsilon(\forall x\,\forall y\,((-\text{ is taller than }-)(x,y))) = T$ iff
      for all γ, $\upsilon_\gamma((-\text{ is taller than }-)(x,y)) = T$
4. Show directly that the following is a tautology.
   $\neg\exists x\,(-\text{ is a wombat})(x) \rightarrow \forall x\,\neg(-\text{ is a wombat})(x)$
5. Classify the following as valid or invalid.
   a. $\forall x\,(-\text{ is a bachelor})(x)$
      Therefore, $\neg((-\text{ is married})(\text{Ralph}))$
   b. $\forall x\,((-\text{ is a dog})(x) \rightarrow (-\text{barks})(x)) \wedge \exists x\,(-\text{ is a dog})(x)$
      Therefore, $\exists x\,((-\text{ is a dog})(x) \wedge (-\text{barks})(x))$
   c. $(-\text{ is a dog})(\text{Ralph}) \vee \neg\exists x\,(-\text{barks})(x)$
      $(-\text{ barks})(\text{Juney})$
      Therefore, $(-\text{ is a dog})(\text{Ralph})$

G. Propositional Logic in Predicate Logic

Example 21 $\forall x\,\forall y\,(((-\text{ is a dog})(x) \vee \neg(-\text{ is a dog})(x)) \vee (-\text{ is a cat})(y))$
Analysis This is true in every model. The part after the quantifiers has form

$(A \lor \neg A) \lor B$, which is a form of a classical propositional logic tautology, and so will be evaluated as true by every valuation.

Propositional forms in predicate logic   A wff A in the language of predicate logic has *propositional form* B where B is a wff of the language of propositional logic, if A is the result of assigning wffs of the language of predicate logic to the propositional variables in B where the *substitution (replacement) is uniform* (the same variable is assigned the same wff throughout B).

Example 22 $\neg((-\text{ is a dog})(x) \land \neg(-\text{ is a dog})(x))$
$\rightarrow \neg((-\text{ is a dog})(x) \land \neg(-\text{ is a dog})(x))$
Analysis   This has propositional form $\neg(p_1 \land \neg p_1) \rightarrow \neg(p_1 \land \neg p_1)$. It also has propositional form $p_2 \rightarrow p_2$.

Example 23 $((-\text{ is a dog})(\text{Ralph}) \rightarrow (-\text{ barks Ralph})) \rightarrow$
$(\neg(-\text{ barks Ralph}) \rightarrow \neg(-\text{ is a dog})(\text{Ralph}))$
Analysis   This has propositional form: $(p_1 \rightarrow p_2) \rightarrow (\neg p_2 \rightarrow \neg p_1)$. That's a propositional tautology. So for every assignment of references σ, if E is the example, $\nu_\sigma(E) = \top$.

Rather than use specific wffs to designate a propositional form, let's use schemes. So we can say that Example 23 has form $(A \rightarrow B) \rightarrow (\neg B \rightarrow \neg A)$. I'll let you show the following.

Theorem 2   If a wff A in the language of predicate logic has propositional form B, and B is a tautology in classical propositional logic, then:
    a. For any assignment of references σ, $\nu_\sigma(A) = \top$.
    b. $\vDash \forall \ldots A$

To invoke Theorem 2 to show that a wff is valid, we'll say *by PC*.

Example 24 $\forall x (-\text{ is a dog})(x) \rightarrow \exists y (-\text{ barks})(y)$
Analysis   This has propositional form $A \rightarrow B$. It is not a tautology of predicate logic.

Example 25 $\forall x (-\text{ is a dog})(x) \rightarrow (-\text{ is a dog})(\text{Ralph})$
Analysis   This also has propositional form $A \rightarrow B$. But it is a tautology of predicate logic.

Example 26 $\exists x_1 (P_3(x_1) \lor \neg P_3(x_1))$
Analysis   Because $P_3(x_1) \lor \neg P_3(x_1)$ has the form of a propositional tautology, in any model it is true of every object in the universe. So $\exists x_1 (P_3(x_1) \lor \neg P_3(x_1))$ is a tautology.

    How did we get into the position of logic guaranteeing the existence of something? The assumption that we are reasoning about things led to understanding the quantifiers in terms of assigning references to variables. That depends on there being at least one object in the universe that variables can take as reference. We have no truth-conditions for a wff of the form $\forall x A(x)$ when the universe is empty. That this example is a tautology just reflects that we are always reasoning about some objects.

G. The Division of Form and Content Verified

The assumption we built into our logic that form and content are distinct should be reflected in our models: If C and D are semantically equivalent, then D can be replace C in any wff that contains C as a part and we get an equivalent wff. To verify that we need to say what we mean by wff being a part of another wff.

Subformulas

If C has length 1, it has one subformula, C itself.

If C has length $n + 1$, then:

    If C has the form $\neg$ A, its subformulas are C, A, and the subformulas of A.

    If C has the form $A \vee B$ or $A \wedge B$ or $A \rightarrow B$, then its subformulas are
        C, A, B, and the subformulas of A and the subformulas of B.

    If has the form $\forall x$ A or $\exists x$ A, then its subformulas are C, A,
        and the subformulas of A.

A is a *proper subformula* of C if A is a subformula of C and A is not C itself.

    We write C(A) to mean that the wff A is a subformula of C.

Example 27 $\exists x \, ((-\text{ is a dog})(x) \wedge (-\text{ barks})(x)) \rightarrow \neg \forall y \, (-\text{ meows})(y)$

Analysis The (abbreviated) subformulas of this are:

    $(-\text{ is a dog})(x)$

    $(-\text{ barks})(x)$

    $(-\text{ meows})(y)$

    $(-\text{ is a dog})(x) \wedge (-\text{ barks})(x)$

    $\exists x \, ((-\text{ is a dog})(x) \wedge (-\text{ barks})(x))$

    $\forall y \, (-\text{ meows})(y)$

    $\neg \forall y \, (-\text{ meows})(y)$

    $\exists x \, ((-\text{ is a dog})(x) \wedge (-\text{ barks})(x)) \rightarrow \neg \forall y \, (-\text{ meows})(y)$

Theorem 3 Substitution of wffs (*the Division of Form and Content verified*)

Let C(A) be a wff in which A is a proper subformula. Let B be a wff with exactly the same variables free as A has. Let C(B) be the result of replacing some but not necessarily all occurrences of A in C by B. In a model, if for every σ, $\vee_\sigma(A) = \vee_\sigma(B)$, then

    a. For every σ, $\vee_\sigma(C(A)) = \vee_\sigma(C(B))$.

    b. $\forall \ldots (A \leftrightarrow B) \rightarrow (C(A) \leftrightarrow C(B))$ is true in the model.

Proof (a) We proceed by induction on the length of C. If C has length 1, then C is A, and we are done. So suppose (a) is true for all wffs of length less than that of C.

If C has the form $\neg$ D, then C(A) is $\neg$ D(A). By induction, for all σ, $\vee_\sigma(C(A)) = \vee_\sigma(C(B))$, and so $\vee_\sigma(\neg C(A)) = \vee_\sigma(\neg C(B))$. I'll leave to you the cases for the other propositional connectives.

Suppose now that C has the form $\forall x\, D$. Then C(A) is $\forall x\, D(A)$. So for any σ,
$$\mathsf{v}_\sigma(\forall x\, D(A)) = \mathsf{T} \quad \text{iff for every } \tau,\, \mathsf{v}_\tau(D(A)) = \mathsf{T}$$
$$\quad\quad\quad\quad \text{iff for every } \tau,\, \mathsf{v}_\tau(D(B)) = \mathsf{T} \text{ by induction}$$
$$\quad\quad\quad\quad \text{iff } \mathsf{v}_\sigma(\forall x\, D(B)) = \mathsf{T}$$

I'll leave to you the case when C has the form $\exists x\, D$. Part (b) then follows. ∎

Aside: *Compound predicates?*
We've been careful to distinguish predicates from other parts of our syntax, particu-larly distinguishing them from terms. We say that "— is a dog" is a predicate, while "(— is a dog) (x_2)" is an atomic wff, made up of a predicate and a term.

But consider:

— is a dog

— is a canine ∧ — is domesticated

It could be argued these are in some sense the same. So shouldn't we consider the latter a predicate, too?

We could say that any open wff is a predicate. So "(— is a dog) (x)" is a predicate, as are:

(— is a dog) (x) ∧ (— barks) (x)
(— is a dog) (x) ∧ (— barks) (y)
(— loves —) (x, x)
(— is a person) (x) ∧ $\forall y\, [\,(-$ is a cat$)\,(y) \rightarrow (-$ hates $-)\,(x, y)\,]$

A compound open wff would then be a compound predicate, while an atomic wff would be an atomic predicate. Atomic predicates would correspond to our simple predicates by dropping the term(s).

The problem with doing this is that we now have:

(— is a dog) (x)
(— is a dog) (y)

Are these the same predicate? If so, then a predicate is not a linguistic unit but has to be something that the linguistic unit stands for. Or we could say that they are not the same but are equivalent. We would need the entire apparatus of classical predicate logic to determine which linguistic expressions are the same predicate or are equivalent predi-cates. We couldn't start our work on classical predicate logic on a clear notion of predicate but would need all of classical predicate logic to yield that.

Key Words

assignment of references	tautology
completeness of the collection of assignments of reference	semantic consequence
	The Semantic Deduction Theorem
valuation	sufficiency of collection of models
extensionality of atomic predications	classical predicate logic
compositionality	propositional form
The Division of Form and Content	by PC
model	subformula

12 Distribution of Quantifiers and Substitution of Variables

    A. Order and Distribution of Quantifiers 93
    Exercises 96
    B. Substitution of Variables 97
    Exercises 100

A. Order and Distribution of Quantifiers

Example 1   All dogs and all cats hate each other.
                All cats and all dogs hate each other.
Analysis Informally these are equivalent because to say "all dogs and all cats" is like picking out any pair. The two quantifiers amount to really just one.

Example 2   Some dog and some cat hate each other.
                Some cat and some dog hate each other.
Analysis These are equivalent because to say "some dog and some cat" is like picking out a pair. The two quantifiers amount to just one.

Theorem 1   For every wff A with exactly n free variables $y_1, \ldots, y_n$,

    $\upsilon(\forall y_1 \ldots \forall y_n A) = T$ iff for every σ, $\upsilon_\sigma(A) = T$.

    $\upsilon(\exists y_1 \ldots \exists y_n A) = T$ iff for some σ, $\upsilon_\sigma(A) = T$.

Proof We prove the first by induction on the number of variables. It's true for $n = 1$ by definition. Suppose now that the theorem is true when there are n variables, and let $x, y_1, \ldots, y_n$ be all the variables free in A .

    $\upsilon(\forall y_1 \ldots \forall y_n \forall x A) = T$
        iff for every σ, $\upsilon_\sigma(\forall x A) = T$ (by induction)
        iff for every σ, for every τ such that $\tau \sim_x \sigma$, $\upsilon_\tau(A) = T$
        iff for every γ, $\upsilon_\gamma(A) = T$ (the collection of assignments of references
                                      is complete)
    The second part follows similarly. ∎

Corollary 2   For every σ, $\upsilon_\sigma(\forall x \forall y A) = \upsilon_\sigma(\forall y \forall x A)$
          $\models \forall \ldots (\forall x \forall y A \leftrightarrow \forall y \forall x A)$

          For every σ, $\upsilon_\sigma(\exists x \exists y A) = \upsilon_\sigma(\exists y \exists x A)$
          $\models \forall \ldots (\exists x \exists y A \leftrightarrow \exists y \exists x A)$

Example 3 $\forall x \exists y (-\text{ is a father of }-)(y, x)$
                $\exists y \forall x (-\text{ is a father of }-)(y, x)$
Analysis For a model with universe all dogs that have ever lived and the usual understanding of "is a father of", the first is true, and the second is false. The order of a pair of different quantifiers does matter.

94   An Introduction to Formal Logic

Our understanding of "all" as "each and every one" and "some" as "at least one" imposes a relation between our use of $\forall$ and $\exists$ in our formal system.

Example 4   Something is a dog.
   Not everything isn't a dog.
Analysis   These are equivalent, and so their formal versions are equivalent:
   "$\exists x\ (x$ is a dog$)$" and "$\neg \forall x \neg\ (x$ is a dog$)$".

Example 5   Everything is a dog.
   There's not something that isn't a dog.
Analysis   These are equivalent., and so their formal versions are equivalent:
   "$\forall x\ (x$ is a dog$)$" and "$\neg \exists x \neg\ (x$ is a dog$)$".

Theorem 3   a. $\models \forall \ldots \forall x\, A(x) \leftrightarrow \neg \exists x \neg\, A(x)$

   b. $\models \forall \ldots \exists x\, A(x) \leftrightarrow \neg \forall x \neg\, A(x)$

   c. $\models \forall \ldots \neg \forall x\, A(x) \leftrightarrow \exists x \neg\, A(x)$

   d. $\models \forall \ldots \neg \exists x\, A(x) \leftrightarrow \forall x \neg\, A(x)$

Proof   Parts (a) and (b) are true because that's how we've chosen to understand $\forall$ in terms of "all" and $\exists$ in terms of "some" in the evaluation in models.
   (c) We need to show that for every assignment of references σ,
   $\nu_\sigma(\neg \forall x\, A(x)) = \mathsf{T}$ iff $\nu_\sigma(\exists x \neg\, A(x)) = \mathsf{T}$.
By (a), for every assignment of references σ, $\nu_\sigma(\ \forall x\, A(x) \leftrightarrow \neg \exists x \neg\, A(x)\ ) = \mathsf{T}$.
So by PC and substitution of wffs (Theorem 3, p. 91)
   $\nu_\sigma(\neg \forall x\, A(x) \leftrightarrow \neg \neg \exists x \neg\, A(x)\,) = \mathsf{T}$.
Hence by PC and substitution, $\nu_\sigma(\neg \forall x\, A(x) \leftrightarrow \exists x \neg\, A(x)\,) = \mathsf{T}$.
   Part (d) follows similarly   ∎

Theorem 4   Distributing $\forall$ across $\to$

Each of the following is a tautology, where x is free in $A(x)$ and $B(x)$ and not free in C.

   a. $\forall \ldots \forall x\, (A(x) \to B(x)) \to (\forall x\, A(x) \to \forall x\, B(x))$

   b. $\forall \ldots (\forall x\, (A(x) \to C) \to (\forall x\, A(x) \to C)$

   c. $\forall \ldots (\forall x\, (C \to B(x)) \to (C \to \forall x\, B(x))$

Proof   a. I'll proceed by showing the contrapositive. Suppose there is some σ such that $\nu_\sigma(\forall x\, A(x) \to \forall x\, B(x)) = \mathsf{F}$. Then $\nu_\sigma(\forall x\, A(x)) = \mathsf{T}$ and $\nu_\sigma(\forall x\, B(x)) = \mathsf{F}$. So there is some $\tau \sim_x \sigma$, such that $\nu_\tau(B(x)) = \mathsf{F}$. But since $\nu_\sigma(\forall x\, A(x)) = \mathsf{T}$, $\nu_\tau(A(x)) = \mathsf{T}$. Hence $\nu_\tau(A(x) \to B(x)) = \mathsf{F}$. And so $\nu_\sigma(\forall x\, (A(x) \to B(x))) = \mathsf{F}$.
   The other parts are proved similarly.   ∎

Theorem 5   Distributions of quantifiers

The following pairs are semantically equivalent, where x is free in $A(x)$ and $B(x)$ and not free in C.

Chapter 12 Distribution of Quantifiers and Substitution of Variables 95

a. $\exists x\, (A(x) \wedge C)$
$\exists x\, A(x) \wedge C$

b. $\exists x\, (C \wedge A(x))$
$C \wedge \exists x\, A(x)$

c. $\forall x\, (A(x) \wedge B(x))$
$\forall x\, A(x) \wedge \forall x\, B(x)$

d. $\forall x\, (A(x) \wedge C)$
$\forall x\, A(x) \wedge C$

e. $\forall x\, (C \wedge A(x))$
$C \wedge \forall x\, A(x)$

f. $\exists x\, (A(x) \vee B(x))$
$\exists x\, A(x) \vee \exists x\, B(x)$

g. $\exists x\, (A(x) \vee C)$
$\exists x\, A(x) \vee C$

h. $\exists x\, (C \vee A(x))$
$C \vee \exists x\, A(x)$

i. $\forall x\, (A(x) \vee C)$
$C \rightarrow \exists x\, A(x)$

j. $\exists x\, (C \rightarrow A(x))$
$C \rightarrow \exists x\, A(x)$

k. $\forall x\, (A(x) \rightarrow C)$
$\exists x\, A(x) \rightarrow C$

l. $\forall x\, (C \rightarrow A(x))$
$C \rightarrow \forall x\, A(x)$

m. $\exists x\, (A(x) \rightarrow C)$
$\forall x\, A(x) \rightarrow C$

n. $\forall x\, (C \vee A(x))$
$C \vee \forall x\, A(x)$

Proof a. For all σ, $\mathsf{v}_\sigma(\exists x\, (A(x) \wedge C)) = T$
  iff there is some τ such that $\tau \sim_x \sigma$ and $\mathsf{v}_\tau(A(x) \wedge C) = T$
  iff there is some τ such that $\tau \sim_x \sigma$ and $\mathsf{v}_\tau(A(x)) = T$ and $\mathsf{v}_\tau(C) = T$
    (since x does not appear in C)
  iff there is some τ such that $\tau \sim_x \sigma$ and $\mathsf{v}_\tau(\exists x\, A(x)) = T$ and $\mathsf{v}_\tau(C) = T$
  iff $\mathsf{v}_\sigma(\exists x\, A(x) \wedge C) = T$.

Part (b) then follows by PC.

c. $\mathsf{v}_\sigma(\forall x\, (A(x) \wedge B(x))) = T$
  iff for every τ such that $\tau \sim_x \sigma$, $\mathsf{v}_\tau(A(x) \wedge B(x)) = T$
  iff for every τ such that $\tau \sim_x \sigma$, $\mathsf{v}_\tau(A(x)) = T$ and $\mathsf{v}_\tau(B(x)) = T$
  iff for every τ such that $\tau \sim_x \sigma$, $\mathsf{v}_\tau(\forall x\, A(x)) = T$ and $\mathsf{v}_\tau(\forall x\, B(x)) = T$
  iff $\mathsf{v}_\sigma(\forall x\, A(x) \wedge \forall x\, B(x)) = T$.

j. $\mathsf{v}_\sigma(\exists x\, (C \rightarrow A(x))) = T$
  iff $\mathsf{v}_\sigma(\neg \forall x \neg (C \rightarrow A(x))) = T$ by Theorem 5
  iff $\mathsf{v}_\sigma(\neg \forall x (C \wedge \neg A(x))) = T$ by PC
  iff $\mathsf{v}_\sigma(\neg (C \wedge \forall x \neg A(x))) = T$ by part (e)
  iff $\mathsf{v}_\sigma(C \rightarrow \neg \forall x \neg A(x)) = T$ by PC
  iff $\mathsf{v}_\sigma(C \rightarrow \exists x\, A(x)) = T$ by Theorem 5 and PC.

k. $\mathsf{v}_\sigma(\forall x\, (A(x) \rightarrow C)) = T$
  iff $\mathsf{v}_\sigma(\neg \exists x \neg (A(x) \rightarrow C)) = T$
  iff $\mathsf{v}_\sigma(\neg \exists x\, (A(x) \wedge \neg C)) = T$
  iff $\mathsf{v}_\sigma(\neg (\exists x A(x) \wedge \neg C)) = T$ by part (d)
  iff $\mathsf{v}_\sigma(\exists x A(x) \rightarrow C) = T$.

I'll leave to you the other parts. ∎

But not all distributions are equivalent.

Example 6 $\exists x (A(x) \wedge B(x))$ is not semantically equivalent to $\exists x A(x) \wedge \exists x B(x)$.
Analysis In a model with universe all living animals:
$\quad \exists x (-$ is a dog$)(x) \wedge \exists x (-$ is a cat$)(x)$ is true
$\quad \exists x ((-$ is a dog$)(x) \wedge (-$ is a cat$)(x))$ is false.
However, the following is a the tautology:
$\quad \exists x (A(x) \wedge B(x)) \rightarrow (\exists x A(x) \wedge \exists x B(x))$

Example 7 $\forall x (A(x) \vee B(x))$ is not semantically equivalent to $\forall x A(x) \vee \forall x B(x)$.
Analysis For a model in which there are only dogs and one fox that barks,
$\quad \forall x ((-$ is a dog$)(x) \vee (-$ barks$)(x))$ is true,
$\quad \forall x (-$ is a dog$)(x) \vee \forall x (-$ barks$)(x))$ is false.

Example 8 $\exists x (A(x) \rightarrow B(x))$ is not semantically equivalent to $\exists x A(x) \rightarrow \exists x B(x)$.
Analysis $\exists x ((-$ is a dog$)(x) \rightarrow (-$ meows$)(x))$ is true of all living canines because we can take x to be assigned some wolf, and then the antecedent is false. But $\exists x (-$ is a dog$)(x) \rightarrow \exists x (-$ meows$)(x))$ is false because the antecedent is true.

Example 9 $\forall x (A(x) \rightarrow B(x))$ is not semantically equivalent to $\forall x A(x) \rightarrow \forall x B(x)$.
Analysis With universe all dogs and cats, $\forall x (-$ is a dog$) \rightarrow \forall x (-$ is a cat$)(x)$ is true because the antecedent is false. But $\forall x ((-$ is a dog$) \rightarrow (-$ is a cat$)(x))$ is false.

Aside What is classical in classical predicate logic
What we've developed here is called *classical* predicate logic because of two features. First, we use classical propositional logic to evaluate the propositional connectives. Second, we take the relation between $\forall$ and $\exists$ as codified in Theorem 3. Both of these points are denied as correct principles of reasoning by some. For example, constructivists deny that $\neg \forall x \neg A(x) \rightarrow \exists x A(x)$ is a tautology. To show that there is a hyena, it's not enough to show that a contradiction follows from assuming everything isn't a hyena: you have to exhibit a hyena to show that "$\exists x (-$ is a hyena$)(x)$" is true. What "exhibit" means will depend on the method for picking out objects from the universe for a realization, quite different for numbers than for hyenas. More generally, constructivists reject the use of *reductio ad absurdum* as a correct way to reason, and with that they deny that $\neg\neg A \rightarrow A$ is a tautology. You can read about such views in Walter Carnielli's and my *Computability*.

Exercises
1. For each of the following, show that it is valid or give a model in which it fails.
    a. $\forall x ((-$ is a dog$)(x) \rightarrow \neg \exists x (-$ is a dog$)(x))$
    b. $\neg \exists x (-$ is a dog$)(x)) \rightarrow \forall x ((-$ is a dog$)(x)$
    c. $\forall y (-$ is a cat$)(y) \rightarrow \neg \exists y \neg (-$ is a cat$)(y))$
    d. $\forall y (-$ is a cat$)(y) \rightarrow \neg \exists y \neg (-$ is a dog$)(y))$

e. $\forall x \forall y [((- \text{ is a dog } -)(x) \land (- \text{ is a cat } -)(x)) \to (- \text{ hates } -)(x,y)]$
$\to \forall y \forall x [((- \text{ is a dog } -)(x) \land (- \text{ is a cat } -)(x)) \to (- \text{ hates } -)(x,y)]$

f. $\forall x \forall y [((- \text{ is a dog } -)(x) \land (- \text{ is a dog } -)(x)) \to (- \text{ hates } -)(x,y)]$
$\to \forall x \forall y [((- \text{ is a dog } -)(x) \land (- \text{ is a dog } -)(x)) \to (- \text{ hates } -)(x,y)]$

g. $\forall x \exists y (- \text{ is heavier than } -)(y,x) \to \exists y \forall x (- \text{ is heavier than } -)(y,x)$

h. $\exists x \forall y (- \text{ is heavier than } -)(y,x) \to \forall y \exists x (- \text{ is heavier than } -)(y,x)$

i. $\forall y (\neg \forall x (- \text{ is heavier than } -)(y,x) \leftrightarrow \exists x \neg (- \text{ is heavier than } -)(y,x)$

2. For each of the following, show that it is valid or give a model in which it fails.

   a. $\forall x [(- \text{ is a cat})(x) \to (- \text{ stinks})(x)]$
   $\to [\forall x (- \text{ is a cat})(x) \to \forall x (- \text{ stinks})(x)]$

   b. $[\forall x (- \text{ is a cat})(x) \to \forall x (- \text{ stinks})(x)]$
   $\to \forall x [(- \text{ is a cat})(x) \to (- \text{ stinks})(x)]$

   c. $\forall x [((- \text{ is a cat})(x) \to (- \text{ stinks})(x)) \land \exists x (- \text{ is a dog})(x)]$
   $\to \forall x [((- \text{ is a cat})(x) \to (- \text{ stinks})(x)] \land \exists x (- \text{ is a dog})(x)$

   d. $\exists x (- \text{ is a dog})(x) \land \exists x (- \text{ stinks})(x)$
   $\to \exists x [(- \text{ is a dog})(x) \land (- \text{ stinks})(x)]$

   e. $\exists x [(- \text{ is a dog})(x) \land \exists y (- \text{stinks})(y)]$
   $\to \exists x (- \text{ is a dog})(x) \land \exists y (- \text{stinks})(y)$

   f. $\exists x ((- \text{ is a man})(x) \lor (- \text{ is a woman})(x)) \to$
   $\exists x (- \text{ is a man})(x) \lor \exists x (- \text{ is a woman})(x)$

   g. $\exists x ((- \text{ is a car})(x) \to (- \text{ is a dog})(x)) \to$
   $(\exists x (- \text{ is a car})(x) \to \exists x (- \text{ is a dog})(x))$

   h. $(\exists x (- \text{ is a dog})(x) \land \exists x (- \text{ barks})(x) \to$
   $\exists x ((- \text{ is a dog})(x) \land (- \text{ barks})(x))$

3. Show the following extension of Theorem 3 of Chapter 11:
   $\vDash \forall \ldots (A \leftrightarrow B) \to (\forall \ldots C(A) \leftrightarrow \forall \ldots C(B))$

4. Prove parts (c), (f), (g), and (h) of Theorem 5.h.

B. Substitution of Variables

There's a lamp on my desk. It doesn't have a name. But I can assert that it's brown by letting "x_1" stand for it and writing:

$(- \text{ is brown})(x_1)$

Equally, I could let "x_{33}" stand for it and assert:

$(- \text{ is brown})(x_{33})$

Both are true or both are false. Any variable is as good as any other as a temporary name in an atomic predication. Equally good is a fixed name for the object since the predicates we use in a semi-formal language are extensional.

What about compound wffs? Here are some examples, where x, y, z, and w are meant to stand for any four distinct variables.

Example 10 $\forall x \, (- \text{ is a dog}) \, (x)$
Analysis This is equivalent to $\forall y \, (- \text{ is a dog}) \, (y)$. And the following is a tautology:
$\quad \forall x \, (- \text{ is a dog}) \, (x) \rightarrow \forall y \, (- \text{ is a dog}) \, (y)$

Example 11 $\forall x \, (- \text{ is a dog}) \, (x) \rightarrow (- \text{ is a dog}) \, (\text{Ralph})$
Analysis This is true in any model. If an open wff, such as "— is a dog", is true of every thing in the universe, then in particular it's true for named things.

Example 12 $\forall x \, (- \text{ is a man}) \, (x) \lor \forall x \, (- \text{ is a woman}) \, (x)$
Analysis This is equivalent to each of the following:
$\quad \forall y \, (- \text{ is a man}) \, (y) \lor \forall y \, (- \text{ is a woman}) \, (y)$
$\quad \forall y \, (- \text{ is a man}) \, (y) \lor \forall x \, (- \text{ is a woman}) \, (x)$
$\quad \forall x \, (- \text{ is a man}) \, (x) \lor \forall y \, (- \text{ is a woman}) \, (y)$
In any model, each is true or each is false.

Example 13 $\forall x \, (\, (- \text{ is a man}) \, (x) \lor (- \text{ is a woman}) \, (x) \,)$
Analysis This is not equivalent to the last example because in the model with universe all living human beings in which we take the "obvious" truth-values it is (arguably) true, and Example 12 is false. But we can use y in place of x: the example is equivalent to
$\quad \forall y \, ((- \text{ is a man}) \, (y) \lor (- \text{ is a woman}) \, (y) \,)$.
But we can't replace just some of the occurrences of x in the example:
$\quad \forall x \, ((- \text{ is a man}) \, (x) \lor (- \text{ is a woman}) \, (y))$ not a closed wff
$\quad \forall y \, ((- \text{ is a man}) \, (x) \lor (- \text{ is a woman}) \, (x))$ not a wff
$\quad \forall x \, ((- \text{ is a man}) \, (y) \lor (- \text{ is a woman}) \, (x))$ not a closed wff
$\quad \forall y \, ((- \text{ is a man}) \, (y) \lor (- \text{ is a woman}) \, (x))$ not a closed wff

Example 14 $\exists x \, \exists y \, (\neg \, (- \text{ lives in the same house as } -) \, (x, y))$
Analysis This is true in a model with universe of all people. It is equivalent to:
$\quad \exists w \, \exists z \, \neg \, (- \text{ lives in the same house as } -) \, (w, z))$
But we can't replace y with x in the example, as these are not even wffs:
$\quad \forall x \, \exists y \, (\neg \, (-\text{lives in the same house as } -) \, (x, x))$
$\quad \forall x \, \exists x \, (\neg \, (-\text{lives in the same house as } -) \, (x, x))$
And if we delete the superfluous quantifiers, we get wffs that are false in that model:
$\quad \forall x \, (\neg \, (- \text{ lives in the same house as } -) \, (x, x))$
$\quad \exists x \, (\neg \, (- \text{ lives in the same house as } -) \, (x, x))$
It's crucial that in the semantic analysis of the original example distinct references can be assigned to x and y in "$(- \text{ lives in the same house as } -) \, (x, y)$".

Example 15 $\forall x \, (\, (- \text{ is a dog}) \, (x) \rightarrow \exists y \, (- \text{ is the father of } -) \, (y, x) \,)$
Analysis This is true in a model where the predicates have their usual meaning and the universe is all dogs that have ever lived. If we replace the second x with y, we get:
$\quad \forall x \, ((- \text{ is a dog}) \, (x) \rightarrow \exists y \, ((- \text{ is the father of } -) \, (y, y)))$

Chapter 12 Distribution of Quantifiers and Substitution of Variables 99

This is false in that model. The problem is that we have replaced an occurrence of x which is bound by $\forall x$ at the beginning of the formula with y which is bound by $\exists y$.

Here are some definitions and notation to help us sort out when we can replace one variable with another in a compound wff.

A term is free for a variable

A(x) means x occurs free in A (other variables may also be free in A).

A term t is *free for an occurrence of x in A* (i.e., free to replace it) iff
  i. The occurrence of x is free.
  ii. If t is y, then the occurrence does not lie within the scope of some occurrence of $\forall y$ or $\exists y$.

The term t is *free for x in A* iff it is free for every free occurrence of x in A.

A(t/x) means the wff that results from *replacing every free occurrence of x* (if any) in A with the term t, unless we specifically say that t replaces some but not necessarily all occurrences of x for which it is free. We call A(t/x) the result of *substituting t for x*.

Example 16 $((-\text{ is a dog})(z) \wedge (-\text{ is a cat})(x)) \rightarrow \neg(-\text{ is the father of }-)(x, y)$
Analysis For this example:
A(x/y) is $((-\text{ is a dog})(z) \wedge (-\text{ is a cat})(x)) \rightarrow \neg(-\text{ is the father of }-)(x, x)$
A(y/x) is $((-\text{ is a dog})(z) \wedge (-\text{ is a cat})(y)) \rightarrow \neg(-\text{ is the father of }-)(y, y)$
A(z/y) is $((-\text{ is a dog})(z) \wedge (-\text{ is a cat})(x)) \rightarrow \neg(-\text{ is the father of }-)(x, z)$
A(Ralph, x) is
  $((-\text{ is a dog})(z) \wedge (-\text{ is a cat})(\text{Ralph})) \rightarrow \neg(-\text{ is the father of }-)(\text{Ralph}, y)$

Example 17 $\forall x (-\text{ is taller than }-)(x, y)$
Analysis In this example:
  y is not free for x because no occurrence of x is free
  x is free for y, and A(x/y) is: $\forall x ((-\text{ is taller than }-)(x, x))$.
  z is free for y, and A(z/y) is: $\forall x ((-\text{ is taller than }-)(x, z))$.
  y is free for y, and A(y/y) is the example itself.
  "Ralph" is free for y, and A(Ralph/y) is: $\forall x ((-\text{ is taller than }-)(x, \text{Ralph}))$

Theorem 7 If y is free for x in A(x), then in any model,
  a. If for every σ, $\vDash_\sigma(A(x)) = T$, then for every σ, $\vDash_\sigma(A(y/x)) = T$.
    $\vDash \forall \ldots \forall x\, A(x) \rightarrow \forall \ldots \forall y\, A(y/x))$
  b. If for some σ, $\vDash_\sigma(A(x))$, then for some σ, $\vDash_\sigma(A(y/x)) = T$.
    $\vDash \forall \ldots \exists x\, A(x) \rightarrow \forall \ldots \exists y\, A(y/x)$

Proof (a) We proceed by induction on the length of A. Suppose that A is a wff of length 1 (atomic) and has form $A(t_1, \ldots, t_k, x, t_{k+1}, \ldots, t_n)$. To prove the contrapositive, suppose that for some σ, $\nu_\sigma(A(y/x)) = F$. Then define τ by $\tau(t_i) = \sigma(t_i)$ for all i, and $\tau(x) = \sigma(y)$. This is possible because x does not appear free in $A(y/x)$ and the collection of assignments of references for the model is complete. By the consistency of predications, $\nu_\tau(A(y/x)) = F$.

Suppose that the theorem is true for all wffs of length $\leq n$ and that A has length n. I'll leave to you the cases when A has the form $B \wedge C, B \vee C, B \rightarrow C$, or $\neg B$. So suppose that A has the form $\forall z\, B$. Then z can't be y, for then y wouldn't be free for x in A. If z is x, then $A(y/x)$ is A, since no occurrence of x in A is free. If z is neither x nor y, and z is free in B, the proof is similar to that for atomic wffs, and I'll leave it to you. The case when A has the form $\exists z\, B$ is done similarly.

The second part follows by Theorem 1.

The proofs for part (b) are similar. ∎

In Theorem 7 we can't claim "iff", as the next example shows.

Example 18 $(-$ is as tall as $-)\,(x, y)$
Analysis In any model in which "is as tall as" has its usual interpretation, the following is true:

If all σ, $\nu_\sigma \vDash (-$ is as tall as $-)\,(x, y)$ then all σ, $\nu_\sigma \vDash (-$ is as tall as $-)\,(y, y)$.

But the converse is false:

If all σ, $\nu_\sigma \vDash (-$ is as tall as $-)\,(y, y)$, then all σ, $\nu_\sigma \vDash (-$ is as tall as $-)\,(x, y)$.

Let x stand for a person who is shorter than the person that y stands for. The problem is that y may appear free in A, so the substitution $A(y/x)$ may introduce new cross-referencing. Avoiding that, we do have equivalences. I'll let you show the following.

Theorem 8 If y is free for x in A, and y is not free in A, then in any model:

  a. $\vDash \forall \ldots \forall x\, A(x) \leftrightarrow \forall \ldots \forall y\, A(y/x))$
  b. $\vDash \forall \ldots \exists x\, A(x) \leftrightarrow \forall \ldots \exists y\, A(y/x))$

Since every name is free for every free variable in a wff, we have the following, which I'll let you prove.

Theorem 9 For every wff $A(x)$ and every name symbol c,

  a. *Existential Generalization* $\vDash \forall \ldots A(c/x) \rightarrow \forall \ldots \exists x\, A(x)$.
  b. *Universal Instantiation* $\vDash \forall \ldots \forall x\, A(x) \rightarrow \forall \ldots A(c/x)$.

The wff $\exists x\, A(x)$ is called an *existential generalization* of $A(c/x)$.
The wff $A(c/x)$ is called an *instantiation* of $\forall x\, A(x)$.

Exercises

In these exercises, x stands for x_1, and y stands for x_2, and z stands for x_3.

1. For each abbreviated wff A below, give $A(y/x)$.

    a. $(-$ is an uncle of $-)\,(x, \text{Ralph})$

b. (— is an uncle of —) (x, y)
c. (— is an uncle of —) (y, x)
d. (— is an uncle of —) (y, Ralph)
e. ∃y ((— is an uncle of —) (y, x))
f. ∀z ((— is an uncle of —) (z, y) → (— is an uncle of —) (x, y))
g. ∀x (— is an uncle of —) (x, y)

2. For each abbreviated wff A below, if y is free for x in A, give A(y/x).
   a. (— is an uncle of —) (x, y) ∨ ∃x (— is a woman) (x)
   b. (— is an uncle) (Ralph) ∨ ∃x (— is a dog) (x)
   c. ∀x ∀y (((— is a dog) (x) ∨ (— is a cat) (y)) → (— barks) (Ralph))
   d. ∃x [(— is in the nucleus of —) (x, y) ∧ (— is an atom) (y)]
      → ∀y (is smaller than —) (x, y)
   e. (— loves —) (x, y) ∨ ∃z ∃y (¬ (— loves —) (y, z))
   f. (— loves —) (x, y) ∨ ∃z ∃x (¬ (— loves —) (z, x)))

3. For each of the following, show that it is valid or give a model in which it fails.
   a. ∀y (— is a cat) (y) → ∀x (— is a cat) (x)
   b. ∀y (— is a cat) (y) → ∃x (— is a cat) (x)
   c. ∀x ∃y (— loves —) (x, y) → ∀x ∃z (x loves z)
   d. ∃x ∀y (— loves —) (x, y) ↔ ∃z ∀y (z loves y)
   e. ∀y ∀x (— loves —) (x, y) → ∀y (y loves y)
   f. (— is a dog) (Ralph) → ∃z (— is a dog) (z)
   g. ∀y ∃x (— loves —) (x, y) → ∃y (y loves y)
   h. ∀y (∃x (— loves —) (x, y) → (y loves y))
   i. ∃x ∀y (— loves —) (x, y) → ∃x (x loves x)
   j. ∃x ∀y (— loves —) (x, y) → ∃y (y loves y)
   k. ∀y (∀x (— loves —) (x, y) → y loves y)

Key Words   order of quantifiers
                  distribution of quantifiers
                  a term is free for a variable
                  substitution of a term for a variable
                  existential generalization
                  universal instantiation

13 An Axiom System for Classical Predicate Logic

I'll set out here an axiom system to characterize the tautologies and valid inferences of classical predicate logic. It is customary to state the axioms schemes of wffs: each wff that is an instance of the scheme is an axiom. I'll write "Axiom" for "axiom scheme".

We use the same notions of proof, syntactic consequence, completeness, consistency, and theory as for propositional logic (p. 43, p. 46, p. 48). Only closed wffs are tautologies or appear in valid inferences, so the axioms of our system must be closed wffs. Since we evaluate the connectives as in classical propositional logic, we can incorporate as axioms the universal closure of any wff that has the form of an axiom of propositional logic.

An axiom system for classical predicate logic

I.   Propositional axioms

$\forall \ldots [\neg A \to (A \to B)]$

$\forall \ldots [B \to (A \to B)]$

$\forall \ldots [(A \to B) \to ((\neg A \to B) \to B)]$

$\forall \ldots [(A \to (B \to C)) \to ((A \to B) \to (A \to C))]$

$\forall \ldots [A \to (B \to (A \wedge B))]$

$\forall \ldots [(A \wedge B) \to A]$

$\forall \ldots [(A \wedge B) \to B]$

$\forall \ldots [A \to (A \vee B)]$

$\forall \ldots [B \to (A \vee B)]$

$\forall \ldots [(A \to C) \to ((B \to C) \to ((A \vee B) \to C))]$

II.   Axioms governing $\forall$

1. If x is free in $A(x)$ and $B(x)$ and not free in C: *distribution of $\forall$*
   a. $\forall \ldots \forall x (A(x) \to B(x)) \to (\forall x A(x) \to \forall x B(x))$
   b. $\forall \ldots (\forall x (A(x) \to C) \to (\forall x A(x) \to C))$
   c. $\forall \ldots (\forall x (C \to B(x)) \to (C \to \forall x B(x)))$

2. $\forall \ldots (\forall x \forall y A \to \forall y \forall x A)$ *commutativity of $\forall$*

3. a. $\forall \ldots (\forall x A(x) \to A(c/x))$ *universal instantiation*
   b. $\forall \ldots (\forall x A(x) \to A(y/x))$

III. Axioms governing the relation between $\forall$ and $\exists$

    4. a. $\forall \ldots (\exists x\, A \to \neg \forall x \neg A)$

       b. $\forall \ldots (\neg \forall x \neg A \to \exists x\, A)$

Rule $\dfrac{A,\ A \to B}{B}$ where A and B are closed formulas   *modus ponens*

Theorem 1   *Soundness of the axiom system for classical predicate logic*
For every closed wff A, if $\vdash A$, then $\vDash A$.
For every closed wff A and every collection of closed wffs Σ, if $\Sigma \vdash A$, then $\Sigma \vDash A$.

Proof   If we can show that each axiom is valid, then, since the single rule of proof preserves truth in a model, each theorem of the system is valid.

    The propositional Axioms are valid by Theorem 2 of Chapter 11 (p. 90).
    Axioms 1 are valid by Theorem 5 of Chapter 12 (p. 94–95).
    Axiom 2 is valid by Corollary 2 of Chapter 12 (p. 93).
    Axiom 3.a is valid by Theorem 9 of Chapter 12 (p. 100).
    Axiom 3.b is valid by Theorem 8 of Chapter 12 (p. 100).
    Axioms 4 are valid by Theorem 3 of Chapter 12 (p. 94).
    Hence, if $\vdash A$, then $\vDash A$.
    Given any collection Σ of closed wffs, we can show that if $\Sigma \vdash A$, then $\Sigma \vDash A$ by restricting to models of Σ. ∎

Theorem 2   For any collection of closed wffs Σ, if $\Sigma \vDash A$ then $\Sigma \vdash A$.

    The proof of this is harder. You can find it in Appendix 4. A crucial step in that proof is showing:

    A collection of wffs Σ is complete and consistent iff
    there is a model M such that Σ is the collection of all wffs true in M.

So a complete and consistent theory is as full a description as possible of a way the world could be, relative to our semantic assumptions and choice of atomic propositions.
    Combining Theorems 1 and 2, we have the following.

Theorem 3   *Completeness of the axiom system of classical predicate logic*
    For any A and any Γ, $\Gamma \vdash A$ iff $\Gamma \vDash A$.

Aside: Truth and provability
For almost all of the history of logic, people thought that there should be one correct system for all reasoning, which they could present with axioms. There was no separation of truth and theorem: what could be proved is true, and what is true could be proved. The only issue was what axioms to take.
    That began to change with the development of non-euclidean geometry in the nineteenth century. The language of geometry could have different models, and what is

true in one, such as Euclid's parallel axiom, need not be true in another. Then in 1931 Kurt Gödel showed that there is a significant difference between truth and provability. He showed that for any axiom system that includes axioms for even a small portion of arithmetic, there are wffs that are true but cannot be proved—unless the system is inconsistent. As a consequence of that proof, there is no mechanical procedure for determining whether wffs of classical predicate logic are valid. You can see Gödel's proof and read about the history of provability and truth in Walter Carnielli's and my book *Computability*.

Key Words   axiom scheme
soundness of an axiom system
completeness of an axiom system

Summary of Chapters 7–13

Let's review what we've done.

Propositional logic is inadequate for formalizing deductions that depend on the internal structure of atomic propositions. But what internal structures?

The subject-predicate distinction in our language is sufficiently central to how we reason that we chose to concentrate on grammatical forms which reflect that distinction. But grammar is not a sufficient guide. Some notion of individual thing is necessary to build upon. We made that explicit in our assumption that the world is made up, at least in part, of individual things and that only propositions about things will be of interest to us in our analyses.

The grammatical categories we chose reflect that view of the world: names, predicates, demonstratives and pronouns, propositional connectives, quantifying phrases, and phrase punctuation. Names we took as we find them in our ordinary language. Then we defined a predicate to be what's left over when the names and pronouns are removed from a proposition. The logical vocabulary consists of variables, formal propositional connectives, and quantifiers. Our goal was to make the syntactic roles of these formal symbols clear and then agree on how we shall understand them.

We codified the syntax of the vocabulary by setting up a formal language. Then we could consider propositions in a semi-formal language, which is what we get by replacing the predicate and name symbols of the formal language with predicates and names from our ordinary language.

For the semantics, we simplified our treatment of names by making the classical abstraction: each name stands for exactly one thing and we take account of no other semantic aspect of a name. Then a predicate is connected to "the world" by putting names, which include variables used as temporary names, in the blanks to get an atomic proposition that is true or false. The truth or falsity of the atomic propositions is taken as given, for it is not for us as logicians to decide on those values. But the classical abstraction of names imposes a requirement for atomic predications: what is true of an object does not depend on what variable or name is used to pick it out.

However, self-referential sentences raised the question whether a coherent definition of truth is possible. To avoid that problem, we chose to exclude such sentences from any semi-formal language.

Reasoning, we know, depends on regularity. As for propositional logic, the truth-value of a proposition depends on its form and the semantic properties of its parts. But here it also depends on what things there are. We agreed that only these factors should be taken into account in our analyses of truth. And rather than leaving implicit what things we are discussing, for each realization we explicitly postulate a universe.

To provide a model for a semi-formal language, we said that we first designate references in the universe for the names. Then we assume a general method for ascribing reference to variables, one that will allow for sufficiently many assignments of references. We then assign truth-values to the atomic propositions and atomic predications relative to those assignments of references. To extend those valuations to all wffs, we adopted an inductive procedure that follows the structure of each wff. That established how we will understand the logical vocabulary. Classical predicate logic, then, is the formal language, notions of realization, of model, of truth in a model, of tautology, and of semantic consequence based on those.

Finally, we saw an axiom system that characterizes syntactically the tautologies and semantic consequence relation of classical predicate logic.

Now it's time to see how to use this logic. In Chapters 14–16 we'll see how to formalize a wide variety of ordinary English propositions and deductions within the framework we've established. We'll also see that much we'd like to analyze remains outside the scope of classical predicate logic.

14 Formalizing in Classical Predicate Logic

    A. Rewriting English Sentences 107
    B. Initial Criteria 107
    C. Common Nouns
           Relative quantification with "all" 108
           Relative quantification with "some" 109
           Common nouns and multiple quantifiers 110
    D. Negations 113
    Exercises 115
    E. Categorematic Words and Meaning Axioms 115
    F. Adjectives and Adverbs 117
    Exercises 119
    G. Mass Terms 120
    H. Time 121
    I. Examples of Formalizing 123
    J. Criteria and Conventions of Formalization 131
    Exercises 135

A. Rewriting English Sentences

We've been very stingy with the grammatical tools we allow ourselves in analyzing propositions. For instance, where is the predicate and where the name in "All dogs bark"? What is the name in "A dog is a dog"? In this chapter we'll see that following the assumptions we used in devising classical predicate logic, we can establish conventions for how to rewrite these and a wide variety of other propositions in such a way that the new English sentence is equivalent to the original for all our logical purposes and can be easily formalized.

    But why should we be concerned with English sentences anymore? It's the vagueness and imprecision of English that drove us to create a formal language. Why not jettison reasoning in English completely and use only semi-formal languages? If we do, then we're committed to reasoning only under the assumptions of classical predicate logic and we will not be able to see the limitations of those assumptions or compare them to other ways of encountering the world.

    I can't give rules for rewriting every proposition we'll want to reason with, and often there's more than one way to rewrite a sentence leading to equally reasonable choices for a formalization. What I'll do in this and the next two chapters is set out many conventions about formalizing sentences in the current tradition of predicate logic.

B. Initial Criteria

To begin, we have the criteria of formalization we adopted with classical propositional logic (pp. 40–41). The first, as modified for predicate logic, is:

Criterion 1 Metaphysics
The formalization respects the assumptions that govern our choice of syncategorematic vocabulary and definition of truth in a model. The constraints we work under when we adopt classical predicate logic must be observed.

The others are:

- If a proposition is informally true due to its form iff formalization is a tautology.

- A proposition is informally false due to its form iff its formalization is an anti-tautology.

- One proposition follows informally from another proposition or collection of other propositions iff its formalization is a formal semantic consequence of the formalizations of the other(s).

These four reflect a deeper criterion that guarantees them.

Criterion 2 Possibilities
In any way the world could be, the proposition and its formalization are both true or both false.

In our formal system, a way the world could be is a model. So we can take account of ways in which the original could be true or could be false only with respect to the assumptions we've built into classical predicate logic. If there is a way the world could be in which a proposition is false yet in all the classical predicate logic models it is true (or vice-versa), then the problem is with the first criterion: some aspect of how we understand the proposition and the world is not compatible with the assumptions of classical predicate logic.

Some examples will make this clearer and help us develop further criteria. When I say in an example that a proposition is true or is false, I mean with respect to a model in which predications are assigned their "obvious" truth-values. When I use x, y, and z, they're meant to stand for x_1, x_2, and x_3. In each analysis, I mark the final choice for what to use as a formalization with (*).

C. Common Nouns

Relative quantification with all

Example 1:   a. *All dogs bark.*
              b. *Ralph is a dog.*
    Therefore,   c. *Ralph barks.*

Analysis The inference is valid, so our formalization should be, too.

The formalizations of (b) and (c) are straightforward. For (a), we follow the analysis with which we began predicate logic (p. 57) and have:

(*)   ∀x (((— is a dog) (x) → (— barks) (x))
       (— is a dog) (Birta)
       Therefore, (— barks) (Birta)

This is valid in classical predicate logic (Universal Instantiation).
   Why not take a realization with universe all dogs and formalize (c) as:

   ∀x (— barks) (x)

But Ralph isn't a dog. We want a formalization that respects Criterion 2 (Possibilities), not a formalization that is good only relative to some models.
   In a model in which there are objects other than dogs in the universe, the quantification in "∀x ((— is a dog) (x) → (— barks) (x))" is restricted to just dogs: we're not asserting that all things bark but only that each dog barks.

Example 2: *All polar bears are white.*
Analysis We can formalize this as we did (a) in Example 1:

(*)   ∀x [(— is a polar bear) (x) → (— is white) (x)]

But what if we have a model with universe all living creatures in Antarctica? There are no polar bears there. So (*) will come out true since the antecedent is always false. But some say that the example would be false. They understand "all" to mean "each and every one and there is at least one". For them, a correct formalization of the example should be:

(x)   ∀x [(— is a polar bear) (x) → (— is white) (x)] ∧ ∃x (— is a polar bear) (x)

   Most logicians nowadays do not want to ascribe an existential assumption to a proposition unless compelled to do so by the role of the proposition in inferences, and they do not consider "All polar bears are white, therefore there is a polar bear" compelling. The general convention now is to use (*) as a formalization of the example, adopting (x) only when the context demands it. This is only a convention. We're not asserting anything about the deep meaning of "all" in English. Nor would a statistical survey of speakers of English be relevant in deciding which convention to adopt, for if there is a disagreement, we need only test which of the formalizations best respects the assumptions of the context in which the proposition is used.

Relative quantification with some

Example 3 *Some cat is feral.*
Analysis We follow the analysis with which we began predicate logic (p. 58):

(*)   ∃x [(— is a cat) (x) ∧ (— is feral) (x)]

   This is also a good formalization of "There is a cat that is feral" and of "There exists a feral cat". That we use the same formalization for each of these as well as the example leads some to say that these sentences all "express the same proposition", but what that proposition is I cannot say.

Example 4: *Some cats are feral.*
Analysis Some say that because of the plural "cats" the example is true only if there is

more than one cat that is feral. We'll see how to formalize that reading in Chapter 16. The tradition now is to formalize this example the same as Example 3 unless context suggests otherwise.

Example 5: *A cat is mewing.*
Analysis For this to be true, there has to be at least one cat that is mewing. So we can rewrite the example as:

There is a cat that is mewing.

Then we can formalize this in the same manner as Example 3:

(*) $\exists x\,[\,(-\text{ is a cat})\,(x) \wedge (-\text{ is mewing})\,(x)\,]$

Example 6: *A dog is a dog.*
Analysis This is surely true due to its form, for we don't need to know anything about dogs to evaluate it. So it should be formalized as a tautology. If we convert the common noun "dog" into a predicate, we get in analogy with our earlier rewritings:

A thing which is a dog is a thing which is a dog.

But we can't proceed as in the last example and formalize this as:

$\exists x\,[\,(-\text{ is a dog})\,(x) \wedge (-\text{ is a dog})\,(x)\,]$

That would be false in a model with universe all cats. Here "a" is meant as "all", the same as in "A dog is a mammal". So we formalize the example not like Example 5 but like Example 1:

(*) $\forall x\,(\,(-\text{ is a dog})\,(x) \rightarrow (-\text{ is a dog})\,(x)\,)$

This is a tautology in classical predicate logic.

Despite the similarity of "a cat is" and "a dog is" in these last two examples, we formalize those phrases differently by reflecting on when the propositions in which they appear would be true. Though we certainly want to formalize propositions of "the same form" in the same way, that constraint must be secondary to ensuring that we have a formalization that will have the same truth-value as the original with respect to any universe.

Some say, though, that the example is not a tautology. For it to be true there has to exist at least one dog, raising the same objections as in Example 2. In that case they should formalize the example as:

$\forall x\,[\,(-\text{ is a dog})\,(x) \rightarrow (-\text{ is a dog})\,(x)\,] \wedge \exists x\,(-\text{ is a dog})\,(x)$

Common nouns and multiple quantifiers

Example 7: *All dogs bark and all cats meow.*
Analysis We can formalize this using the method of Example 1 for each part:
(a) $\forall x\,[\,(-\text{ is a dog})\,(x) \rightarrow (-\text{ barks})\,(x)\,]$
         $\wedge\ \forall y\,[\,(-\text{ is a cat})\,(y) \rightarrow (-\text{ meows})\,(y)\,]$

This is equivalent to (Theorem 6.e of Chapter 12):
(b) $\forall x\,\forall y\,[\,(\,(-\text{ is a dog})\,(x) \rightarrow (-\text{ barks})\,(x)\,)$
         $\wedge\ (\,(-\text{ is a cat})\,(y) \rightarrow (-\text{ meows})\,(y)\,)\,]$

And it is also equivalent to both:

(c) $\forall x\, [\,(-\text{ is a dog})\,(x) \to (-\text{ barks})\,(x)\,[$
     $\wedge\ \forall x\, [\,(-\text{ is a cat})\,(x) \to (-\text{ meows})\,(x)\,]$

(d) $\forall x\, [\,(\,(-\text{ is a dog})\,(x) \to (-\text{ barks})\,(x)\,)$
     $\wedge\ (\,(-\text{ is a cat})\,(x) \to (-\text{ meows})\,(x)\,)\,]$

(Theorem 7 of Chapter 12 and Theorem 5 of Chapter 12)

     I think that (a) and (b) are preferable because the different variables will lead to less confusion. And then (a) is preferable as part of a general method of formalizing parts in accord with what we have already agreed on.

(*) $\forall x\, [\,(-\text{ is a dog})\,(x) \to (-\text{ barks})\,(x)\,]$
     $\wedge\ \forall y\, [\,(-\text{ is a cat})\,(y) \to (-\text{ meows})\,(y)\,]$

This choice illustrates a general convention that lets us approach formalizing in an inductive manner.

> *Convention A* (*Parts to Whole*)   We formalize parts of propositions in accord with our established agreements whenever possible, then formalize the way those parts are put together in the original.

Example 8:   *There is a dog that meows and a cat that barks.*
Analysis   We formalize in accord with Convention A and what we did in Example 3, using different variables as in Example 7:

(*) $\exists x\,(\,(-\text{ is a dog})\,(x) \to (-\text{ meows})\,(x)\,)$
     $\wedge\ \exists y\,(\,(-\text{ is a cat})\,(y) \to (-\text{ barks})\,(y)\,)$

     Though the following is equivalent (Theorem 5, Chapter 12), it is not in accord with Convention A:

$\exists x\, \exists y\, [\,(\,(-\text{ is a dog})\,(x) \to (-\text{ meows})\,(x)\,)$
$\wedge\ (\,(-\text{ is a cat})\,(y) \to (-\text{ barks})\,(y)\,)\,]$

Example 9:   *Some dog barks and every cat meows.*
Analysis   We formalize in accord with Convention A:

(*) $\exists x\,(\,(-\text{ is a dog})\,(x) \wedge (-\text{ barks})\,(x)\,)$
     $\wedge\ \forall y\,(\,(-\text{ is a cat})\,(y) \to (-\text{ meows})\,(y)\,)$

Example 10:   *Every dog hates every cat.*
Analysis   We have a relation "— hates —". As in Example 1, it's not all objects that are relevant for the first blank in this but only those that are dogs, and it's not all objects that are relevant for the second blank but only those that are cats. We need both these restrictions as antecedents, and since no order is intended or needed for the pair of universal quantifiers, we can put them both at the start:

(a) $\forall x\, \forall y\, [\,(\,(-\text{ is a dog})\,(x) \wedge (-\text{ is a cat})\,(y)\,) \to (-\text{ hates }-)\,(x, y)\,]$

This and the example are both true or both false relative to any model.

Alternatively, we could follow the earlier examples by taking care of one quantifier at a time:

(b) $\forall x \,[\,(\,(-\text{ is a dog})\,(x) \wedge \forall y\,(\,(-\text{ is a cat})\,(y)\,) \rightarrow (-\text{ hates }-)\,(x, y)\,]$

This is equivalent to (a) (Theorem 5 of Chapter 12) and puts the formalization of the "every" for cats later in the proposition as in the original. There is no principled choice to be made here. But if we are to have some regularity in formalizing, some rules that we can use as a guide rather than improvising with each new example, we need to choose one for our standard method. Before we do that, let's look at other pairs of quantifiers.

Example 11: *Every dog hates some cat.*
Analysis In accord with what we did above, we have three choices:

(a) $\forall x \,\exists y\,[\,(\,(-\text{ is a dog})\,(x) \wedge (-\text{ is a cat})\,(y)\,) \rightarrow (-\text{ hates }-)\,(x, y)\,]$
(b) $\forall x \,\exists y\,[\,(\,(-\text{ is a dog})\,(x) \rightarrow (\,(-\text{ is a cat})\,(y) \wedge (-\text{ hates }-)\,(x, y)\,)\,]$
(c) $\forall x \,[\,(-\text{ is a dog})\,(x) \rightarrow \exists y\,(\,(-\text{ is a cat})\,(y) \wedge (-\text{ hates }-)\,(x, y)\,)\,]$

These are semantically equivalent, as you can show, and each is true in a model iff for that universe the example is true.

Example 12: *Some dog hates all cats.*
Analysis Again we have three semantically equivalent choices for formalizing this in accord with what we've seen so far:

(a) $\exists x \,\forall y\,[\,(\,(-\text{ is a dog})\,(x) \wedge (-\text{ is a cat})\,(y)\,) \rightarrow (-\text{ hates }-)\,(x, y)\,]$
(b) $\exists x \,\forall y\,[\,(\,(-\text{ is a dog})\,(x) \rightarrow (\,(-\text{ is a cat})\,(y) \wedge (-\text{ hates }-)\,(x, y)\,)\,]$
(c) $\exists x \,[\,(-\text{ is a dog})\,(x) \rightarrow \forall y\,(\,(-\text{ is a cat})\,(y) \wedge (-\text{ hates }-)\,(x, y)\,)\,]$

As a general method, putting all the quantifiers at the start, giving the quantifier for each conversion maximal scope still leaves us choices: (a) or (b) in Example 11, and (a) or (b) in Example 12. There is little to motivate a choice between those. However, making the conversions in order, one at a time, with the quantifiers for the conversions having minimal scope, does determine the formalizations: (b) in Example 10, (c) in Example 11, and (c) in Example 12. This gives us a clear inductive method, building up from parts. So let's opt for this latter method. Thus, we have for Example 10:

(*) $\forall x \,[(-\text{ is a dog})\,(x) \wedge \forall y\,(\,(-\text{ is a cat})\,(y)\,) \rightarrow (-\text{ hates }-)\,(x, y)\,)\,]$

for Example 11:

(*) $\forall x \,[\,(-\text{ is a dog})\,(x) \rightarrow \exists y\,(\,(-\text{ is a cat})\,(y) \wedge (-\text{ hates }-)\,(x, y)\,)\,]$

and for Example 12:

(*) $\exists x \,[\,(-\text{ is a dog})\,(x) \rightarrow \forall y\,(\,(-\text{ is a cat})\,(y) \wedge (-\text{ hates }-)\,(x, y)\,)\,]$

In sum, we adopt the following convention in accord with Convention A (Parts to Whole).

Convention B (*Converting Common Nouns into Predicates*)
To formalize a proposition containing a common noun (phrase) used as a subject or object, we convert the noun (phrase) into a predicate and supply some form of quantification.

Universal quantification over just those objects covered by a common noun is modeled by taking the converted predicate as the antecedent of a conditional that is universally quantified.

Existential quantification is formalized by taking the predicate as a conjunct.

When there is more than one common noun in the proposition, we apply these rules sequentially, using minimal scope for each quantification.

For the cases we've seen, this method can be expressed rather succinctly.

Let γ and δ stand for common or collective nouns, and $\gamma(-), \delta(-)$ stand for the conversions of those into predicates. We formalize:

All γ are δ.	$\forall x \, (\gamma(x) \rightarrow \delta(x))$
Some γ are δ.	$\exists x \, (\gamma(x) \wedge \delta(x))$
All γ are R to all δ.	$\forall x \, (\gamma(x) \rightarrow \forall y \, (\delta(y) \rightarrow R(x,y)))$
Some γ is R to some δ.	$\exists x \, (\gamma(x) \wedge \exists y \, (\delta(y) \wedge R(x,y)))$
All γ is R to some δ.	$\forall x \, (\gamma(x) \rightarrow \exists y \, (\delta(y) \wedge R(x,y)))$
Some γ is R to all δ.	$\exists x \, (\gamma(x) \wedge \forall y \, (\delta(y) \rightarrow R(x,y)))$

These are conventions, not criteria of formalization, guides for regular formalizations of common nouns in accord with our criteria of formalization. With these we can formalize propositions with any number of common nouns, converting them into predicates in succession.

D. Negations

Example 13: *Cows are not white.*
Analysis Both the example and "All cows are white" are false. So we can't formalize the example as:

$\neg \, \forall x \, [\, (- \text{ is a cow}) \, (x) \rightarrow (- \text{ is white}) \, (x) \,]$

The relative quantification in the example applies to all that follows, which includes the negation applied to the predicate. We formalize the example:

(*) $\quad \forall x \, [\, (- \text{ is a cow}) \, (x) \rightarrow \neg \, (- \text{ is white}) \, (x) \,]$

Example 14: *Every dog doesn't meow.*
Analysis What does every dog do? It doesn't meow. We formalize this in accord with what we did in the last example:

(*) $\forall x\,((-\text{ is a dog})(x) \to \neg(-\text{ meows})(x))$

The negation applies to the last predication, not to the entire proposition. This is equivalent to both:

$\neg \exists x \neg\,((-\text{ is a dog})(x) \to \neg(-\text{ meows})(x))$

$\neg \exists x\,((-\text{ is a dog})(x) \wedge (-\text{ meows})(x))$

(Theorem 4 of Chapter 12 and by PC). But we choose (*) because it places the negation as in the original.

Convention C (*Negations*)
We take a negation to apply to an entire proposition or wff only if it expressly occurs that way in the proposition we are formalizing. In all other cases, we give wider scope to the quantification we use to convert a noun into a predicate than to the negation.

This is part of a more general convention we've been using implicitly, which we singled out for propositional logic (Example 11 of Chapter 5, p. 38).

Convention D (*Grammar*)
The grammar of the original is preserved by the formalization. That is, the structure of the original proposition with respect to the grammatical categories we have assumed (names, predicates, connectives, quantifiers, variables, phrase markers) is respected by the formalization. This includes respecting the order of the parts.

Example 15: *Zélia is unmarried.*
Analysis If we take "— is unmarried" as a predicate, then it and "(— is married)" will have no connection. But we do recognize some connection, deeming valid:

(a)   Zélia is unmarried.
      Therefore, it's not the case that Zélia is married.

I take the example to be equivalent to:

(b)   Zélia is not married.

Then we can formalize the example as:

(*) $\neg\,(-\text{ is married})(\text{Zélia})$

But doesn't (*) violate Convention C (Negations)? There is no other place for the negation to go in formalizing (a), for we cannot split "is" and "married". Our conventions are secondary for we must first satisfy the criteria, in this case respecting the validity of (a). This is part of a general convention.

Convention E (*Parity of Form*)
A regular translation of certain words or parts of words as syncategorematic terms is observed. Further formalizations are regular in the sense that each proceeds in analogy with agreed-upon formalizations of others.

For example, we formalize "not" as ¬, "all" as ∀, "every" as ∀, "some" as ∃, "there exists" as ∃, and "it" as a variable, "un-" as negation, so long as doing so doesn't contravene the criteria of formalization.

Exercises
1. Why formalize ordinary language propositions instead of reasoning entirely in semi-formal languages?
2. For each of the following, formalize it or explain why it is not formalizable.
   a. All wombats squeak.
   b. At least one wombat squeaks.
   c. At least two wombats squeak.
   d. If Ralph is a dog, then some puppet is a dog.
   e. Dick is taller than Zoe.
   f. Every person is taller than Spot.
   g. Every man is taller than some woman.
   h. Every person knows some person who is rich.
   i. Some clowns are unhappy.
   j. A dog is a canine.
   k. A dog is unhappy.
   l. Zeke disrespected Zoe.

E. Categorematic Words and Meaning Axioms

Example 16 *Dick is a bachelor.*
  Therefore, *Dick is a man.*
Analysis The obvious formalization is:

(a) (— is a bachelor) (Dick)
     Therefore, (— is a man) (Dick)

This is invalid: we could have a model in which "— is a bachelor" is true only of rabbits. Yet the example is valid. But it's not valid due to its form, which is all we can hope to respect in our formalizations. It is valid due to the meaning of the words.

If "— is married" is also in a semi-formal language, we could formalize the example as:

  (— is a man) (Dick) ∧ ¬ (— is married) (Dick)
  Therefore, (— is a man) (Dick)

That's valid. But this would be an analysis before formalization, replacing one categorematic word (or phrase) with another. That's not the work of the logician. It's for a linguist or a sociologist to tell us these relations, or for us as knowledgeable speakers of English. If we want to respect that meaning, we should make it explicit:

(b) ∀x [(— is a bachelor) (x) ↔ ((— is a man) (x) ∧ ¬ (— is married) (x))]

Then the formalization (a) is acceptable (to us as English speakers) only in models in which (b) is true. Then (b) is a good formalization *relative to the meaning axiom*:

(*)   (— is a bachelor) (Dick)
      Therefore, (— is a man) (Dick)
      relative to $\forall x \, [\, (— \text{ is a bachelor}) (x) \leftrightarrow ((— \text{ is a man}) (x) \land \neg (— \text{ is married}) (x)) \,]$

A *formalization* is a translation into a semi-formal language that sticks as closely as possible to the form and words of the original, as understood by our agreements establishing predicate logic. An *analysis* is an attempt to make explicit our understandings of the words in a proposition that are not covered by the agreements establishing predicate logic and our conventions for formalizing. Incorporating analyses of words into the process of formalization would change the project of logic from investigation of logical validity based on form to one based on the meaning of categorematic words.

Criterion 3   Categorematic Words
The formalization contains exactly the same categorematic words as the original, allowing for changes of grammar to satisfy the other criteria.

To the extent that an analysis can be captured by one or many semi-formal propositions, we can use those *meaning axioms* to supplement a formalization, saying that the formalization is good only relative to our formalized meaning assumption(s).

We need to allow changes of grammar to formalize "dogs" as "— is a dog".

Criterion 3 along with the criterion of respecting possibilities is the best we can do for formalizing the meaning of a categorematic word, relative to other categorematic words.

Example 17: *Zeke disrespected Zoe*.
Analysis Proceeding as in Example 15, we get:
      $\neg (— \text{ respected } —) (\text{Zeke, Zoe})$
If we think that "disrespected" means more) than just "not respected", we should take "— disrespected —" as a predicate and add a meaning axiom:
      $\forall x \, \forall y \, [\, (— \text{ disrespected } —) (x, y) \to \neg (— \text{ respected } —) (x, y) \,]$

Example 18   Through any two points there is a line.
Analysis In studying geometry we don't already know what the words "point", "line", and "parallel" mean. Rather, we try to come to some understanding of those by giving a formal analysis to guide us in using those notions in our reasoning. We set out axioms that serve as a guide for how to reason about points and lines. For Euclidean geometry we take "point" and the relation "— lies between — and —" as primitive and circumscribe their meaning with axioms (you can see how that's done in my *Classical Mathematical Logic*). Then anything that satisfies those axioms will serve as a point, whether it be dots drawn on a piece of paper or places marked by a surveyor's stakes. That is not analysis before formalization but an attempt to explicate or come to agreement on meaning via a formal theory.

Example 19: *7 is less than 12*.

Analysis We understand "is less than" well enough when we say "Suzy weighs less than Tom". But just as with geometry, there are different ways to understand it for numbers, people, and more. So we set up a formal theory of orderings, replacing "is less than" with a formal symbol, "<" with axioms governing how it is to be used:

$\forall x \, \forall y \, [\, (-<-)(x, y) \to \neg\, (-<-)(y, x)\,]$

$\forall x \, \neg\, (-<-)(x, x)$

$\forall x \, \forall y \, [\,(\,(-<-)(x, y) \land (-<-)(y, z)\,) \to (-<-)(x, z)\,]$

This is not analysis before formalization but the establishing of a theory of orderings. We adopt a convention to govern formalizations such as this and the previous example.

> *Convention F* (*Formal Theories*)
> If we give an analysis of a particular predicate by formalizing it with a formal symbol whose meaning we stipulate either by semantic agreements or an axiomatization, we may choose in advance to recognize certain other predicates or grammatical variations as being formalizable by the same symbol.

For example, we might use the same symbol in geometry to formalize both "(— lies on —) (x, y)" and "(— passes through —) (y, x)".

F. Adjectives and Adverbs

Example 20: *Ralph is a purple dog*.
  Therefore, *Ralph is purple*.

Analysis This is a valid inference. And its formalization is valid, too:

(*)   (— is purple) (Ralph) ∧ (— is a dog) (Ralph)
     Therefore, (— is purple) (Ralph)

Example 21: *Bon Bon is a small donkey*.
  Therefore, *Bon Bon is small*.

Analysis This is not a valid inference. The premise is true and the conclusion is false: my donkey Bon Bon is huge compared to a mouse or even a cocker spaniel.

  The adjective "small" is a *relative adjective*, that is, it can be evaluated only relative to a kind of thing: a small elephant, a small mouse, a small skyscraper. There is no absolute idea of what's small. So we can't detach the adjective from the noun to formalize the example as:

  (— is small) (Bon Bon) ∧ (— is a donkey) (Bon Bon)

We have to take "(— is a small donkey)" as a simple predicate. But then we can't respect that the following inference is valid:

(a)   Bon Bon is a small donkey.
     Therefore, Bon Bon is a donkey.

We could add an *ad hoc* assumption:

(b) $\forall x \, [\, (-\text{ is a small donkey})(x) \to (-\text{ is a donkey})(x)\,]$

Then we'd formalize (a) only relative to (b). But the following is a valid inference, too:

Tico is a small dog.
Therefore, Tico is a dog.

So we'd have to add an axiom governing "small dog". And we'd also have to add axioms governing "small building" and "small rock". None of these would be a meaning axiom; only the collection of them might serve to show how to use "small", at least relative to the predicates in the semi-formal language. And then we'd have to set out axioms to respect that the following are valid:

Bidú is a big dog.
Therefore, Bidú is big.

Birta is an old dog.
Therefore, Birta is a dog

Yet it's clear that each of these is valid due to its form.

The problem is we haven't taken account of such a form with the grammatical categories we adopted for classical predicate logic. The example does not contravene Criterion 1 (Metaphysics); we just don't have the tools to formalize it. We can extend classical predicate logic to formalize uses of relative adjectives by taking account of the internal structure of predicates, viewing, for example, "— is a small donkey" as a simple predicate plus a modifier: "(— is a donkey)/small". You can see how to do that in my *The Internal Structure of Predicates and Names*.

In contrast, in the previous example "purple" is an *absolute adjective*: it doesn't matter what kind of thing we call purple as the same standard is used for all things. So it can be detached and used as a predicate.

Example 22: *Zoe is a beautiful woman.*
Analysis There has been a long debate about whether "beautiful" is absolute or relative. If relative, we can't formalize the example with the tools we have. If absolute, we have:

$(-$ is beautiful$)$ (Zoe) $\wedge$ $(-$ is a woman$)$ (Zoe)

Example 23: *Juney is barking loudly.*
     Therefore, *Juney is barking*.
Analysis We can't detach "loudly" from "barking" because the adverb "loudly" describes barking, not Juney: "Juney is barking and Juney is loudly" is nonsense.

Adverbs modify verbs, and whatever verbs describe in the world, it's not individual things. In classical predicate logic, the premise of this example has to be formalized as atomic. To respect the validity of the example, we'd have to add an axiom "(— is barking loudly) (Juney) $\rightarrow$ (— is barking) (Juney)", or more generally "$\forall x$ [(— is barking loudly) (x) $\rightarrow$ (— is barking) (x)]". We'd also have to add axioms to ensure the validity of:

Dick is eating quickly.
Therefore, Dick is eating.

Tom ran fast.
Therefore, Tom ran.

Yet it seems these are all valid due to their form.

We can formalize inferences that involve adverbs in the same way as inferences that involve relative adjectives by reading, for example, the predicate "— is barking loudly" as a simple predicate plus a modifier "(— is barking)/loudly". That's also in *The Internal Structure of Predicates and Names*.

Example 24: *Every dog which yelps also barks.*
Analysis The phrase "which yelps" is a subordinate adjectival clause. We can rewrite the example as:

(a)   Every thing which is a dog which yelps is also one that barks.

By deleting "also" we get what I think is an equivalent proposition, so we can ignore that word. We don't want to take "— is a dog which yelps" as a predicate because then we couldn't respect the validity of "Ralph is a dog which yelps, therefore Ralph is a dog". So we formalize (a) as:

(*) $\forall x [((- \text{ is a dog}) (x) \wedge (- \text{ yelps}) (x)) \to (- \text{ barks}) (x)]$

The adjectival clause is added as a conjunct to the predicate that formalizes the common noun (phrase).

Example 25: *Every dog which yelps or whimpers also barks.*
Analysis I'll let you provide the justification for formalizing this as:

(*) $\forall x [(- \text{ is a dog}) (x) \wedge ((- \text{ yelps}) (x) \vee (- \text{ whimpers}) (x)) \to (- \text{ barks}) (x)]$

The conjunct used to formalize an adjectival clause need not be atomic.

Example 26: *Every person who owns a puppy is not depressed.*
Analysis We can formalize this as:

$\forall x [((- \text{ is a person}) (x) \wedge \exists y ((- \text{ owns}) (x, y) \wedge (- \text{ is a puppy}) (y)))$
$\to \neg (- \text{ is depressed}) (x)]$

We're not restricted to only propositional connectives in formalizing an adjectival clause.

Exercises

1. Why require that the proposition we're formalizing and the formalization have exactly the same categorematic words?
2. Formulate a general convention for formalizing adjectival clauses.
3. What is a meaning axiom? Why do we use meaning axioms for formalizing?
4. Why can't we formalize the use of a relative adjective in a proposition?
5. Why can't we formalize the use of an adverb in a proposition?
6. For each of the following, formalize it or explain why it is not formalizable.
    a. Harold is a smart sheep.
    b. Every horse that neighs is healthy.
    c. No person knows a green sheep.
    d. Some dog that is owned by a woman barks at all mailmen.

G. Mass Terms

Example 27:   a. Snow is white.
               b. All that's white is not black.
   Therefore, c. Snow is not black.

Analysis The inference is valid. Can we formalize it in classical predicate logic?

In (a) and (c) we seem to be using "Snow" as a name. But what thing does it name? What individual is there that is snow? Snow is physical, but it's not in one place. Yes, snow is distinct from all else in the world, but we can't point to it, even in theory. We can only use it to describe lots of masses or things that are made of snow. I can point and say, "That ball is made of snow". I can point to a tree when walking and say, "The stuff on that tree is snow". But I can't point to snow complete, all snow.

Don't we have the same problem with "Dogs"? Consider:

(d)   Dogs bark.
      All that barks is not a cat.
      Therefore: Dogs are not cats.

We recognize this as valid, and "Dogs" is used as a subject much like a name. We cannot point to dogs complete at one time. I can point to this dog or that dog, but not all dogs even in theory. Rather, "Dogs" describes lots of individual things we call "dogs". So we reformulate "Dogs" as a predicate and formalize (d) as:

(e) $\forall x\,((-\text{ is a dog})(x) \to (-\text{ barks})(x))$
      $\forall x\,((-\text{ barks})(x) \to \neg(-\text{ is a cat})(x))$
      Therefore, $\forall x\,((-\text{ is a dog})(x) \to \neg(-\text{ is a cat})(x))$

This is valid. And in any model in which "dog", "cat", and "barks" are given their usual interpretation, (e) reflects what we intended in (d).

Why can't we do the same with snow? We might try formalizing the example as:

$\forall x\,((-\text{ is snow})(x) \to (-\text{ is white})(x))$
$\forall x\,((-\text{ is white})(x) \to \neg(-\text{ is black})(x))$
Therefore, $\forall x\,((-\text{ is snow})(x) \to \neg(-\text{ is black})(x))$

But what could be used as a reference for x to make "$(-\text{ is snow})(x)$" true? What could we put into the universe of a model that is all snow? We would have to include all quantities of snow, whether a snowball, or a snowman, or what's lying on the branches of a tree. We would also have to include all parts of those, for every part of snow is snow. I can pick up a handful of snow and there are innumerable "quantities" of snow in it. Only with a description picking out a particular quantity of snow can we specify a thing that is snow because snow, unlike dogs, does not come in identifiable quantities, it does not come in distinct parts each of which is snow. Snow is a mass, not a collection of things. We have no conception of how we could pick out in a general way just any item that "snow" describes.

What holds for "snow" holds equally for other *mass terms* that we use to describe masses: "gold", "mud", "water", ... Every part of a mass is a mass of the same kind. In contrast, no part of a thing is a thing of the same kind. No part of my dog Birta is a dog. Masses are not like things, and talk about masses cannot be reduced to talk about things. We cannot formalize propositions that involve mass terms because there is a metaphysical mismatch and so Criterion 1 (Metaphysics) would be violated. Masses and unidentifiable bits of masses are not things nor collections of things.

It would seem that we could formalize some uses of mass terms where a particular object or kind of object made up of that mass is meant, with predicates like "(— is a snowball)", "(— is a snow man)" "(— is a snow drift)". But we would have no way to recognize that the categorematic word "snow" appears in each of those, no way to relate those predicates. At best, we could add a predicate "— is made of snow".

Example 28: *Running is fun.*
Analysis Running not a thing, not an individual thing that we can pick out and re-identify. We use "running" to describe a process, and like masses, every part of running is running, and there is no smallest bit of running. We cannot formalize a proposition that involves a word meant to describe a process, for to do so would contravene Criterion 1.

H. Time

Example 29: a. *All dogs bark.*
  b. *Birta is a dog.*
 Therefore, c. *Birta barks.*
Analysis This is a valid inference, and we formalize it:

(*) $\forall x (((-\text{ is a dog}) (x) \rightarrow (-\text{ barks}) (x))$
  (— is a dog) (Birta)
  Therefore, (— barks) (Birta)

The formal inference is valid, too. We justify that by saying that if a thing is a dog then it barks, so if Birta is a dog, she barks. It is remarkably unclear what we mean by this. The reading of the predicates isn't atemporal, since Birta is a thing in time. Nor is the reading of the predicates omnitemporal, since we don't mean that Birta is barking all the time. It's more like ascribing essential attributes or permanent capabilities or dispositions to things. If Birta is a dog, that is, if she has that attribute without any reference to time, then Birta barks, without any reference to time. But that Birta barks is not an essential attribute of Birta. Or perhaps we need to think it is in order to use classical predicate logic to formalize the example. And perhaps that is indeed what we mean by the example. This is how we must understand the wffs at (*): they are true because being a dog or barking is an attribute we ascribe to an object independent of time.

Example 30: *Every puppy barks.*
Analysis We formalize this:

(*) $\forall x (((-\text{ is a puppy}) (x) \rightarrow (-\text{ barks}) (x))$

Suppose we want to formalize both this example and the previous example in a semi-formal language. The predicates "— is a puppy" and "— is a dog" are related in meaning: a puppy is an immature dog. So we adopt an axiom to codify that meaning:

(a) $\forall x (((- \text{ is a puppy}) (x) \leftrightarrow ((- \text{ is a dog}) (x) \wedge \neg (- \text{ is mature}) (x)))$

What does this mean? It's said to be atemporal, but nothing is a puppy atemporally. Suppose that my dog Birta is in the universe of a model. Is she classified as a mature dog or as a puppy? To use (a), we must choose, but that means we can formalize no true proposition about Birta when she was a puppy.

Whatever properties we ascribe to an object in a model must be unchanging. That's just to say that the truth-values of predications in a model are fixed. That does not preclude, however, having all of the following true in a model:

$(-$ was a puppy$)$ (Birta)

$(-$ is a dog$)$ (Birta)

$(-$ is mature$)$ (Birta)

Even with the axiom at (a), these are not inconsistent. Such a model amounts to setting out what is true of certain objects at one particular time. In 2009 both of the following are true:

$(-$ is a dog$)$ (Birta) $\wedge$ $(-$ is mature$)$ (Birta)

$(-$ was a puppy$)$ (Birta)

But if that's how we interpret our models, then (a) does not ensure that the following is true in that model:

$(-$ was a dog$)$ (Birta) $\wedge \neg (-$ was mature$)$ (Birta)

We would have to add as well:

$\forall x [(- \text{ was a puppy}) (x) \leftrightarrow \neg ((- \text{ was a dog}) (x) \wedge (- \text{ was mature}) (x))]$

But that's false, for Birta was both a puppy and was a mature dog, just not at the same time. We'd need to find a better way to relate "— was a dog" to "— is a dog," and "— was a woman" to "— is a woman," and .... To reason about things in time in classical predicate logic, about how objects have different properties at different times, we would have to adopt meaning axioms governing every predicate.

In classical predicate logic no account is taken of the times at which the things in the universe of a model are meant to exist. They all exist in a timeless status: the coming into existence and going out of existence of them is of no concern. The only existence we reason about in classical predicate logic is that which is suitable to allow for a thing to be the value given to a variable by an assignment of references.

Reference in classical predicate logic is timeless. We say that we can assign Socrates to x, though Socrates does not exist now. This is not the issue of whether we have the ability to pick out a thing, for we think that we can pick out Socrates among all other things to be the reference of a variable.

Quantification is timeless, too. For an existential quantification to be true, there must be something in the universe that can be assigned to the variable that makes the resulting proposition true, and that assignment is timeless. For a

Chapter 14 Formalizing in Classical Predicate Logic

universal quantification to be true, each thing in the universe, regardless of any considerations of time, must satisfy the predicate.

Classical predicate logic is useful for formalizing reasoning about things outside of time, or about essential attributes or permanent capacities of things that are in time. Any reasoning that depends essentially on when things exist or when a predicate is true of a thing is outside the scope of what we have done here. But that is not because of a metaphysical mismatch. We simply haven't included grammatical categories such as tenses or time markers in our vocabulary that would allow us to take account of time when reasoning about things. You can see how to extend classical predicate logic to formalize reasoning that takes account of time in my *Time and Space in Formal Logic*.

I. Examples of Formalizing

Let's look at some examples to see how to use the criteria and conventions we have adopted.

Example 31: *Every counting number is even or odd.*
Analysis The word "odd" is defined as "not even" whenever it's used with counting numbers. So it's no analysis before formalization to use that definition.

(*) $\forall x\, [\,(-\text{ is a counting number})\,(x) \rightarrow ((-\text{ is even})\,(x) \vee \neg(-\text{ is even})\,(x))\,]$

Note how putting the disjunction in a clause resolves the problem we had with formalizing this example in propositional logic in Example 24 of Chapter 5.

Example 32: *All dogs bark.*
         Therefore, *some dog barks*.
Analysis The formalization is straightforward using our conventions on converting nouns into predicates:

(*) $\forall x\, [\,(-\text{ is a dog})\,(x) \rightarrow (-\text{ barks})\,(x)\,]$
      Therefore, $\exists x\, [\,(-\text{ is a dog})\,(x) \wedge (-\text{ barks})\,(x)\,]$

This is not valid because we agreed to interpret "all" as "each and every one (and there need not be one)". We could have a model in which there are no dogs, so the premise would be true but conclusion false.

If you think the example is valid, you could formalize it as the valid inference:

$\forall x\, [\,(-\text{ is a dog})\,(x) \rightarrow (-\text{ barks})\,(x)\,] \wedge \exists x\, (-\text{ is a dog})\,(x)$
Therefore, $\exists x\, [\,(-\text{ is a dog})\,(x) \wedge (-\text{ barks})\,(x)\,]$

Example 33: *Pegasus is a horse.*
         Therefore, *there exists a horse*.
Analysis The obvious formalization is:

   $(-\text{ is a horse})\,(\text{Pegasus})$
   Therefore, $\exists x\, (-\text{ is a horse})\,(x)$

That's valid, an example of existential generalization. But Pegasus is a mythical creature, a winged horse. Still, if we want to use it as a name, it has to pick out something in the universe. So we can't use this formalization and the example is unformalizable. In *The Internal Structure of Predicates and Names* you can see how to extend classical predicate logic to allow for reasoning with names that do not pick out anything.

Example 34: *If Ralph is a dog, then anything is a dog.*
Analysis The formalization is straightforward, formalizing "any" as ∀.

(*)   (— is a dog) (Ralph) → ∀x (— is a dog) (x)

Example 35: *If anything is a dog, then Ralph is a dog.*
Analysis Parity of form with the previous example must be overruled in favor of the natural reading of "any" as "some" in this example.

(*)   ∃x (— is a dog) (x) → (— is a dog) (Ralph)

Stephen C. Kleene, in *Mathematical Logic*, 1967 (pp. 140–147), suggests we read:
  If S, then anything P   as   S → ∀x P
  If anything S, then P   as   ∃x S → P

He gives a careful and extensive discussion of the formalization of propositions involving "all", "any", "some", and other words we might wish to read as ∀ or ∃.

Example 36: *If something weighs over 30 lbs., it cannot be sent by parcel post.*
Analysis We can't invariably formalize uses of "some" as ∃. The example is a law, a regulation applying to all things, not to at least one. So we formalize the example:

(*)   ∀x [(— weighs over — pounds) (x, 30) → ¬ (— can be sent by parcel post)]

I've taken "30" as a name, thinking of the proposition used along with propositions involving other weights.

Example 37: *Everyone has seen Marilyn Monroe on TV.*
Analysis It would not be correct to formalize "everyone" as we formalize "everything": the example is not false just because some cat has not seen Marilyn Monroe on TV. In English, "one" in "everyone" indicates we are talking about people, just as it does in "anyone", "someone", and "no one".

(*)   ∀x [(— is a person) (x) → ((— has seen on TV —) (x, Marilyn Monroe))]

> *Convention G* ("*anyone*" *and* "*anybody*")
> We can rewrite propositions in which "everyone", "anyone" "some one", and "no one" occur by replacing "one" by "thing which is a person" without violating Criterion 3 (Categorematic Words).
>   The same rewriting applies to "body" in "everybody", "anybody", "somebody", and "nobody".

Example 38: *There is a nice cat.*
Analysis There are at least three ways we could construe this example depending on the context in which it is uttered.
    Zoe might have said this in response to an assertion such as "All cats are nasty", using "there is a" as "there exists at least one". This we can formalize as:

$\exists x \,[\, (-\text{ is a cat})\,(x) \wedge (-\text{ is nice})\,(x) \,]$

Or Zoe might utter the sentence when pointing to some cat, "Look! There is a nice cat." If by "there" Zoe means to refer to a location, the proposition will have to remain outside the scope of predicate logic unless we can replace the indexical with a name of a place that we can agree to view as an object.

Or Zoe might use "there" as a way to point to a specific cat that has a name. If she were pointing to Puff we could rewrite the proposition as "Puff is a nice cat", which we could formalize as "$(-\text{ is a cat})\,(\text{Puff}) \wedge (-\text{ is nice})\,(\text{Puff})$".

None of this applies, however, if we think that "nice" is not an absolute adjective, which I think is right, for the standard for a cat to be nice is much lower than for a supermarket clerk to be nice.

Example 39: *Spot is owned by Dick.*
  Therefore, *Dick owns something.*
Analysis This is clearly valid. Yet if we formalize the inference directly we have:

  $(-\text{ is owned by }-)\,(\text{Spot}, \text{Dick})$
  Therefore, $\exists x\,(-\text{ owns }-)\,(\text{Dick}, x)$

This is invalid, since "$(-\text{ is owned by }-)$" and "$(-\text{ owns }-)$" are used as predicates and hence have no relation. So this formalization would contravene Criterion 2 (Possibilities).

But the premise is equivalent to:

  Dick owns Spot.

We can rewrite a proposition in passive form by converting it into active form, reversing the subject and object roles. Doing that, we can formalize the example:

(*) $(-\text{ owns }-)\,(\text{Dick}, \text{Spot})$
  Therefore, $\exists x\,(-\text{ owns }-)\,(\text{Dick}, x)$

This is valid (existential generalization).

Example 40: *Spot is owned.*
  Therefore, *Something owns Spot.*
Analysis The premise is in passive form. But it has no indirect object. In English, though, it is equivalent to:

(a) Spot is owned by something.

Any proposition in passive form is equivalent to one in which a neutral indirect object is supplied: "something". So we can formalize this example as:

(*) $\exists x\,(-\text{ owns }-)\,(\text{Dick}, \text{Spot})$
  Therefore, $\exists x\,(-\text{ owns }-)\,(\text{Dick}, \text{Spot})$

Supplying the indirect object in this case is just recognizing that the inference is valid.

Example 41: *Someone has climbed Mt. Everest.*
  Therefore, *Mt. Everest has been climbed.*
Analysis The conclusion is equivalent to "Mt. Everest has been climbed by something", which in turn is equivalent to "Something has climbed Mt. Everest". So we can formalize the example as: :

(*) $\exists x\,[\,(-\text{ is a person})\,(x)\,\wedge\,(-\text{ has climbed }-)\,(x, \text{Mt. Everest})\,]$
Therefore, $\exists x\,(-\text{ has climbed }-)\,(x, \text{Mt. Everest})$

This is valid, as you can show.

Example 42: *Mr. Everest has been climbed.*
  Therefore, *Someone has climbed Mt. Everest.*
Analysis We can formalize this using what we did in the last example and Convention G ("anyone"):

(*) $\exists x\,(-\text{ has climbed }-)\,(x, \text{Mt. Everest})$
Therefore, $\exists x\,[\,(-\text{ is a person})\,(x)\,\wedge\,(-\text{ has climbed }-)\,(x, \text{Mt. Everest})\,]$

This is not valid: we could have a model in which the only creature to have climbed Mt. Everest is an abominable snowman. The indirect object we supply is neutral with respect to what kind of thing.

These conversions of passive into active do not depend on the particular categorematic word. They are based on the assumptions built into our (English) grammar. So we adopt a convention, not a criterion, to govern them.

  Convention H (*Passive into Active*)
  Any proposition in passive form is equivalent to one in active form.
  If the proposition has an indirect object, then the active form reverses
  the role of subject and object. If the proposition has no indirect object,
  a neutral one is first supplied: "by something".

Example 43: *Some dog barks at all women and men.*
Analysis The formalization is routine because we've established conventions on converting nouns into predicates in an inductive order (p. 113).

(*) $\exists x\,[\,(-\text{ is a dog})\,(x)\,\wedge\,\forall y\,(\,(-\text{ is a woman})\,(y)\,\rightarrow$
  $\forall z\,(\,(-\text{ is a man})\,(z)\,\rightarrow\,(\,(-\text{ barks at }-)\,(x, y)\,\wedge\,(-\text{ barks at }-)\,(x, z)))\,]$

Without those conventions we could work a long time deciding how to formalize this.
  I've taken "barks at" as an predicate here to illustrate how to convert many nouns in one proposition into predicates. That should be supplemented with a meaning axiom: $\forall x\,\forall y\,(\,(-\text{ barks at }-)\,(x, z)\,\rightarrow\,(-\text{ barks})\,(x)\,)$.

Example 44 *There is a woman who is a sister of a man who hates all cats.*
Analysis Again, the formalization is routine using our conventions.

(*) $\exists x\,[\,(-\text{ is a woman})\,(x)\,\wedge\,\exists y\,(\,(-\text{ is a sister of }-)\,(x, y)\,\wedge\,(-\text{ is a man})\,(y)$
  $\wedge\,\forall z\,(\,(-\text{ is a cat})\,(z)\,\rightarrow\,(-\text{ hates }-)\,(y, z)\,)\,]$

Example 45 *Horses eat apples.*
Analysis Here the simple present is used for what is usual, or normal, or habitual rather than as always. Which of those uses is intended is sometimes difficult to gauge. Is "Juney barks at night" to be taken in the sense of "always" or in the sense of habitual? In most cases it doesn't matter, since we can retain the simple present in the semi-formal

wff, assimilating all uses of the simple present to a timeless verb, as in "Dogs bark". But for this example there is no sense of a normal or usual if in the formalization we relate one horse to one apple:

$$\forall x\,((-\text{ is a horse})\,(x) \to \forall y\,((-\text{ is an apple})\,(y) \to (-\text{ eats } -)\,(x, y)))$$

Indexing the use of the simple present as, say, "eats$_{\text{habitual}}$" in the formalization, would not restore a normal or habitual sense of "eats" to a relation of a particular horse eating a particular apple. So I suggest that we amplify the tense of the verb to make explicit what I understand to be its sense of a normal or usual action.

(*) $\quad \forall x\,[\,(-\text{ is a horse})\,(x) \to \forall y\,((-\text{ is an apple})\,(y)$
$\quad\quad \to (-\text{ has the capacity and preference to eat } -)\,(x, y))\,]$

This is to take "has the capacity and preference to eat" as an indivisible phrase replacing "eats" that does not introduce any new categorematic word. This is the best we can do without taking tenses and time into account.

Example 46: *A horse will eat an apple.*
Analysis Without contextual clues, I view the example as a general statement about horses, a natural law, reading "a" as "any" and treating the verb as timeless. So we can formalize this as the last example.

Example 47: *A horse will eat an apple, but not an orange.*
Analysis I take "but not an orange" as elliptical for "but a horse will not eat an orange", and formalize the example in accord with the last example:

(*) $\quad \forall x\,[\,(-\text{ is a horse})\,(x) \to \forall y\,((-\text{ is an apple})\,(y) \to$
$\quad\quad (-\text{ has the capacity and preference to eat } -)\,(x, y))\,] \land$
$\quad \forall x\,[\,(-\text{ is a horse})\,(x) \to \forall y\,((-\text{ is an orange})\,(y) \to$
$\quad\quad \neg\,(-\text{ has the capacity and preference to eat } -)\,(x, y))\,]$

This formalization follows from Convention A (Parts to Whole), Convention C (Negations), and Convention B (Converting Common Nouns into Predicates).

We could, however, rewrite the example as:

A horse will eat an apple, but it will not eat an orange.

In that case the relative quantification for "a horse" will extend over the whole proposition, and we'd get:

(a) $\quad \forall x\,[\,(-\text{ is a horse})\,(x) \to$
$\quad\quad \forall y\,((-\text{ is an apple})\,(y) \to (-\text{ has the capacity and preference to eat } -)\,(x, y)) \land$
$\quad\quad \forall y\,((-\text{ is an orange})\,(z) \to \neg\,(-\text{ has the capacity and preference to eat } -)\,(x, z))\,]$

The choice between (a) and (*) depends on how we expand "but not an orange". Since the two English readings are informally equivalent, the two semi-formal readings should be equivalent, and they are (Theorem 5 of Chapter 12). I think (*) is better because it preserves the apparent grammar of the example as joining two propositions.

Example 48: *A horse will eat an apple, but not a dog.*
Analysis It would be odd if not disingenuous to take "but not a dog" as elliptical for "but a horse will not eat a dog". The word "but" suggests some surprise, yet everyone

knows that horses don't eat dogs. Rewriting the example as "A horse will eat an apple, but a dog will not eat an apple", the formalization is straightforward according to our conventions.

(*) $\forall x$ [($-$is horse) (x) $\wedge$ $\forall y$ (($-$ is an apple) (y) $\rightarrow$ ($-$ has the capacity
    and preference to eat $-$) (x, y)] $\wedge$
   $\forall z$ [($-$ is a dog) (z) $\rightarrow$ $\forall y$ (($-$ is an apple) (y) $\rightarrow$ ¬ ($-$ has the capacity
    and preference to eat $-$) (z, y)]

Syntax alone is not a sufficient guide for formalizing. We must take into account how we understand the proposition.

Example 49:   *Dogs have eaten apples.*
Analysis   Here we have no law, no general statement about all dogs, but an observation that sometime in the past some dog(s) has eaten some apple(s). We must override the convention for formalizing "dogs".

(*) $\exists x$ [($-$ is a dog) (x) $\wedge$ $\exists y$ [($-$ is an apple) (y) $\rightarrow$ ($-$ has eaten $-$) (x, y)]

This can be good only if we are formalizing for a model of what is true at some specific time.

   In this example, it does seem that "dogs" means more than one, but to account for that in our formalization, we'll have to wait for Chapter 16.

Example 50:   *Dick has a dog.*
   Therefore, *Dick has something.*
Analysis   The deduction is informally valid. So "has" must be construed as a binary predicate. There are other uses of "has" that cannot be assimilated to this sense of it as "possesses". So let's index "has" with the letter "p" and formalize by our convention for converting nouns into predicates.

(*) $\exists x$ [($-$ has$_p$ $-$) (Dick, x) $\wedge$ ($-$ is a dog) (x)]

Example 51:   *Dick doesn't have any cats.*
Analysis   I understand "any" as "a" in the sense of "at least one" and rewrite the example as:

   Dick doesn't have a cat.

I take the negation to apply to the proposition as a whole and rewrite further:

   Not: Dick has a cat.

The formalization then follows by our conventions for converting nouns into predicates and the discussion of "has" in the last example:

   ¬$\exists x$ [($-$ is a cat) (x) $\wedge$ ($-$ has$_p$ $-$) (Dick, x)]

Or we could enforce a reading of "any" as $\forall$ and formalize this example as:

   $\forall x$ [($-$ is a cat) (x) $\rightarrow$ ¬ ($-$ has$_p$ $-$) (Dick, x)]

We have a choice: shall we stress the relation of this example to "Dick has a dog" or shall we follow our agreement to formalize "any" as $\forall$? The two formalizations are semantically equivalent, as you can show.

Example 52: *Dick has a cold.*
  Therefore, *Dick has something.*
Analysis This inference has the same grammatical form as the last example, but is invalid. A cold is not a thing. Compare: "has the flu", "has a backache"', "has measles". In these, "has" is not meant in the sense of "possesses an object" but is part of a unary predicate indicating that the subject satisfies some condition. In the conclusion, though, "has" must be taken to be binary in the sense of "possesses" as in the last two examples So we formalize this example as:

(*)   (— has a cold) (Dick)
      Therefore, $\exists x [(— \text{has}_p —) (\text{Dick}, x)]$

The criterion of parity of form is violated in order to respect Criterion A (Metaphysics).
  The formalization is invalid, but some say that the example is informally valid because it has the same form as the previous example. To make that claim we must either view a cold as a thing or else eschew any notion of thing from our logic and rely entirely on syntax.

Example 53: *A cat is never your friend.*
Analysis Is this a proposition? Aren't the words "never" and "your" too vague? Perhaps it's just an exclamation. I think it's a proposition because when I say it, people disagree, which shows that they've accepted it as a proposition.
  I understand the example as a statement of a natural law, akin to "Electrons have spin" or "Atoms never disintegrate". In that sense, it's atemporal. So "never" can be read as simply "not". The pronoun "your" I understand to introduce generality, as in "anyone's", rather than referring to the person to whom the example might be addressed. Note that I've taken "your" as "anyone's" not "anything's", generalizing the example to all people. So rewriting, we have:

  A cat is not a friend of any person.

The first "a" should be taken in the sense of "all" if the example is to be understood as a natural law. Then we can formalize the example as:

(*) $\forall x [(— \text{is a cat}) (x) \rightarrow \forall y ((— \text{is a person}) (y) \rightarrow \neg (— \text{is a friend of} —) (x, y))]$

Example 54: *Prince Albert married Queen Victoria.*
  Queen Victoria is a head of state.
  Therefore, *Prince Albert married a head of state.*
Analysis The inference is valid. The formalization of it in accord with our convention of converting nouns into predicates is valid, too:

(*)   (— married —) (Prince Albert, Queen Victoria)
      (— is a head of state) (Queen Victoria)
      Therefore, $\exists x ((— \text{married} —) (\text{Prince Albert}, x) \land (— \text{is a head of state}) (x))$

Example 55: *Prince Albert married Queen Victoria.*
  Queen Victoria is a widow.
  Therefore, *Prince Albert married a widow.*
Analysis This argument has the same form as the previous one, but it's invalid. In Example 54, any way in which the premises are true, which includes the times they

describe, the conclusion is, too. But in this example, the premises are true of different times. We have no way to take account of that difference in classical predicate logic. So the example is not formalizable.

Again, grammatical form in English is not a sure guide to formalizing.

Example 56: Dick: I can't find my glasses. They've got to be somewhere.
Zoe: Duh. *Everything is somewhere.*
Analysis Zoe's assertion limits her conception of things to those that exist in locations, eliminating talk of abstract things. But to formalize what she says, we'd have to take locations to be things and quantify over those. That assumption is not easy to integrate into classical predicate logic; you can see how in my *Time and Space in Formal Logic*.

Example 57: *Dick and Tom lifted the table.*
Analysis It would be wrong to formalize the example as:

(a)   (— lifted the table) (Dick) ∧ (— lifted the table) (Tom)

That could be true if Dick lifted it at one time and Tom at another, while the example means that they lifted the table together. The "and" in the example isn't used as a propositional connective but as a conjunction of terms. We can extend predicate logic to allow for formalization of conjunctions of terms, as you can read in my *The Internal Structure of Predicates and Names*.

Example 58: *Whatever swims has gills.*
Analysis The word "whatever" is used for generality in the sense of "all". We can rewrite the example as:

Everything that swims has gills.

We can formalize this as:

(*) $\forall x\,((-\text{ swims})(x) \rightarrow (-\text{ has gills})(x))$

I'll leave to you to formalize the example viewing a gill as a thing.

Example 59: *Whatever Ralph did yesterday, he is doing again today.*
Analysis If we rewrite this example in parity with the previous one, we get:

Everything that Ralph did yesterday, he is doing again today.

This would be to treat what verbs stand for—processes or actions—as things that we can quantify over. But doings, processes, or actions are not obviously things. We don't take a proposition and remove the verbs and call that a predicate. Trying to force action-talk into thing-talk violates Criterion A (Metaphysics). The example cannot be formalized in classical predicate logic.

Example 60: *Gold is more valuable than iron.*
Analysis This is a proposition, and one that's important in our daily lives. But we can't formalize it because it uses mass terms. Nor can we reduce it to talk of bits of gold and bits of iron, for it's not equivalent to "Every bit of gold is more valuable than every bit of iron": a trainload of iron will fetch more on the market than a wedding ring.

Example 61:   Barking is in a dog's nature.
Analysis This proposition, too, contains a mass term: "barking". So it's not formalizable unless we were to assimilate it to "All dogs bark", which does not seem equivalent.

Example 62:   Marilyn Monroe had all the qualities of a great actress.
Analysis In classical predicate logic we quantify over things. Are qualities things? Are "beautiful", "sexy", "able to portray emotions" things? A quality is something a thing has or doesn't have. It's like a property, such as the property of being blue.

In classical predicate logic, we can say that a thing possess wisdom by saying that "— is wise" is true of it. We can say that a thing has the quality of fatherhood by saying that "— is a father" is true of it. To talk of all or some qualities or properties is to talk of all or some predicates. We have no way to do that in our semi-formal languages, though we do talk that way in our reasoning about our formal systems. It's possible to extend classical predicate logic to allow for quantifying over predicates, as you can see in my *Classical Mathematical Logic*.

Example 63:   It's raining.
Analysis This is a proposition, assuming we agree on the time and place at which it is written or uttered. But it's not formalizable, for what objects is it about? Perhaps we could say that "it" refers to the weather, as in "The weather is raining". But "weather" is a mass term, or at least not an individual thing, and the weather doesn't rain: it's rainy. Perhaps we should rewrite the proposition as "Rain falls"? But that introduces a new categorematic word, and "rain" is a mass term, too.

To formalize this we need an entirely different kind of formal logic., which you can see in my *Reasoning about the World as Process*.

J. Criteria and Conventions of Formalization

When we formalize, we are translating from our ordinary language to a semi-formal language. Our goal, as in all translation, is to preserve meaning. But a formalization can respect only what we have agreed to pay attention to in classical predicate logic, and that is truth and falsity and naming. Overarching all of our work at preserving meaning is our fundamental assumption that the world is made up of things: meaning that does not fit into that view cannot be recognized in classical predicate logic.

We take into account not just whether a proposition, such as "Ralph is a dog" or "All dogs bark", is true or false, but the ways in which it could be true or could be false. We do that with models in which we stipulate how we understand the connectives, the variables with their assignments of references, and the quantifiers. Within any one model, we give the meaning of the categorematic words by stipulating which atomic propositions they appear in are true. The inferential relations in which the categorematic words appear we can track by looking at all or some specified class of models.

Three criteria are fundamental in ensuring that a formalization is good given this analysis of meaning and structure.

Criterion 1 Metaphysics
The formalization respects the assumptions that govern our choice of syncategorematic vocabulary and definition of truth in a model. The constraints we work under when we adopt classical predicate logic must be observed.

Criterion 2 Possibilities
In any way the world could be, the proposition and its formalization are both true or both false.

Criterion 3 Categorematic Words
The formalization contains exactly the same categorematic words as the original, allowing for changes of grammar to satisfy the other criteria.

To the extent that an analysis can be captured by one or many semi-formal propositions, we can use those meaning axioms to supplement a formalization, saying that the formalization is good only relative to our formalized meaning assumption(s).

Criterion 2 is meant to ensure that formalizations preserve the logical relations we find in ordinary reasoning:

- A proposition is informally true due to its form iff its formalization is a tautology.
- A proposition is informally false due to its form iff its formalization is an anti-tautology.
- One proposition follows informally from another proposition or collection of other propositions iff its formalization is a formal semantic consequence of the formalizations of the other(s).

For a formalization to be good, it must satisfy these criteria. But often there are equally good choices for how to formalize an ordinary language proposition. So we adopt conventions as guides to using these criteria.

First, we want to make our formalizations regular in order to build up a general method of formalizing.

Convention: *Parity of form*
Formalizations should be regular in the sense that each proceeds in analogy with agreed-upon formalizations of others.

A regular translation of certain words or parts of words as syncategorematic terms is observed.

Partly we do this by respecting the grammar of the propositions we formalize.

Convention: *Grammar*
The grammar of the original is preserved by the formalization. That is, the structure of the original proposition with respect to

the grammatical categories we have assumed (names, predicates, connectives, quantifiers, variables, phrase markers) is respected by the formalization. This includes respecting the order of the parts.

But we respect the grammar only if it does not contravene our criteria. If grammar were so regular that we could use it as a guide, and if we could always respect the convention of parity of form, we would not need to use semi-formal languages. We are uncovering the implicit assumptions of the grammar of English propositions when we formalize them. At best, we can hope to agree to some conventions with clear enough exceptions and then follow those when we can.

One way we try to follow grammar is to build up formalizations inductively.

Convention: *Parts to whole*
We formalize parts of propositions in accord with our established agreements whenever possible, then formalize the way those parts are put together in the original.

One grammatical category which is not recognized in classical predicate logic is so closely tied to our conception of the world as made up of things that we can and should devise a convention governing it.

Convention: *Converting common nouns into predicates*
To formalize a proposition containing a common noun (phrase) used as a subject or object, we convert the noun (phrase) into a predicate and supply some form of quantification.

Universal quantification over just those objects covered by a common noun is modeled by taking the converted predicate as the antecedent of a conditional that is universally quantified.

Existential quantification is formalized by taking the predicate as a conjunct.

When there is more than one common noun in the proposition, we apply these rules sequentially, using minimal scope for each quantification.

Let γ and δ stand for common or collective nouns, and $\gamma(-)$, $\delta(-)$ stand for the conversions of those into predicates. We formalize:

All γ are δ. $\forall x\,(\gamma(x) \to \delta(x))$

Some γ are δ. $\exists x\,(\gamma(x) \land \delta(x))$

All γ are R to all δ. $\forall x\,(\gamma(x) \to \forall y\,(\delta(y) \to R(x,y)))$

Some γ is R to some δ. $\exists x\,(\gamma(x) \land \exists y\,(\delta(y) \land R(x,y)))$

All γ is R to some δ. $\forall x\,(\gamma(x) \to \exists y\,(\delta(y) \land R(x,y)))$

Some γ is R to all δ. $\exists x\,(\gamma(x) \land \forall y\,(\delta(y) \to R(x,y)))$

It's difficult to find a general rule for how we use negations in English. At best we have the following.

Convention: Negations
We take a negation to apply to an entire proposition or wff only if it expressly occurs that way in the proposition we are formalizing. In all other cases, we give wider scope to the quantification we use to convert a noun into a predicate than to the negation.

Another grammatical construction not directly taken into account in classical predicate logic is the passive-active distinction.

Convention: Passive into active
Any proposition in passive form is equivalent to one in active form. If the proposition has an indirect object, then the active form reverses the role of subject and object. If the proposition has no indirect object, a neutral one is first supplied: "by something".

Not grammar but conventional use of English is recognized by the following convention.

Convention: "anyone" and "anybody"
We can rewrite propositions in which "everyone", "anyone" "some one", and "no one" occur by replacing "one" by "thing which is a person" without violating Criterion 3 (Categorematic Words).
    The same rewriting applies to "body" in "everybody", "anybody", "somebody", and "nobody".

Another way we treat meaning is to allow for formalizations that are meant to make explicit how we shall use certain informal predicates.

Convention: Formal theories
If we give an analysis of a particular predicate by formalizing it with a formal symbol whose meaning we stipulate either by semantic agreements or an axiomatization, we may choose in advance to recognize certain other predicates or grammatical variations as being formalizable by the same symbol.

In the examples, we saw that there are many kinds of propositions we cannot formalize:

- We cannot formalize reasoning that involves relative adjectives.
- We cannot formalize reasoning that involves adverbs.
- We cannot formalize reasoning that depends on time.

Such reasoning is not in conflict with the metaphysics of classical predicate logic. We just haven't accommodated such reasoning with the grammatical categories we start with here.

We also saw:

- We cannot formalize reasoning that involves mass terms or process words.

Such reasoning does conflict with the metaphysics of classical predicate logic, for we cannot construe such reasoning as being about individual things.

Exercises

1. Show that the following are equivalent:
   $\forall x\,((-\text{ barks})\,(x) \rightarrow (-\text{ is a dog})\,(x))$
   $\neg\,\exists x\,[\,(\neg\,(-\text{ is a dog})\,(x)) \land (-\text{ barks})\,(x)\,]$

2. Formalize the following or explain why no good formalization is possible.
   a. Every bear growls.
   b. Every bear growls or roars.
   c. All dogs that are unfriendly are wild.
   d. A scout is reverent.
   e. A lady is present.
   f. Dick gave Spot a bone.
   g. Dick gave Spot some food.
   h. Ralph was snoring heavily.
   i. Ralph's snoring disturbed me.
   j. Alcoholics cannot drive a car.
   k. Scholars who have dogs do not own cats.
   l. Hungry dogs that chase cats never catch them.
   m. Cats that are owned by women eat mice.
   n. Someone who doesn't like dogs will not be a postman.
   o. It's snowing.
   p. Paris is in France.
   q. $2 + 2 = 4$
   r. Seven plus eight is fifteen.
   s. Dick is strong and handsome.
   t. Dick is a man and a student.
   u. Dick dislikes Puff.
   v. If even one teacher disliked doing homework, then every student would, too.
   w. If a person sees more than one dog, then they will run quickly.
   x. There is a round square.

y. Atoms have diameter less than 1/2 cm.
z. Every man amuses himself. (Give a general rule for reflexives.)
aa. God helps those who help themselves.
bb. Everyone who takes Math 101 will fail or pass.
cc. If Trump did everything he was accused of doing, he deserved to go to jail.
dd. Bananas are cheaper by the dozen.
ee. Suzy even loves Spot.
ff. Suzy loves even Spot.
gg. Every pack of dogs has a leader.
hh. Something exists.

3. Formalize "There is a cow that is not white" (compare Example 13).

4. Formalize the following and say whether the inference is valid or invalid, or explain why no good formalization is possible.

   a. Every man will be saved or damned.
      Therefore, every man will be saved or every man will be damned.
      (Due to Ockham)

   b. Horses are mammals.
      Therefore, something is a mammal.

   c. Some teachers are nasty.
      Some nasty teachers yell.
      All nasty teachers who yell yell at students.
      Therefore, some teachers yell at students.

   d. If strawberries are red, then some color-blind people cannot see strawberries among their leaves.
      Strawberries are red.
      Therefore, some color-blind people cannot see strawberries among their leaves.
      (Due to Frege)

   e. Some vegetables are turnips.
      Everyone who likes fruit hates turnips.
      No one both likes and hates anything.
      Therefore, some vegetables are not fruits.

   f. The just is fair.
      The fair is good.
      Therefore, the just is good.
      (Due to Sextus Empiricus)

   g. Suzy is a student of Dr. E.
      Therefore, Suzy is a student.

   h. Every child loves his mother.
      Every mother has a child.
      Therefore, every mother is loved.\

i. Any fool can solve Exercise 1.
Dick cannot solve Exercise 1.
Therefore, Dick is not a fool.
(This and (j)–(l) are variations on exercises in Kleene's *Mathematical Logic*.)

j. If anyone can solve Exercise 1, some mathematician can solve it.
Reginald is a mathematician and cannot solve Exercise 1.
Therefore, Exercise 1 cannot be solved.

k. Any mathematician can solve Exercise 1 if anyone can.
Reginald is a mathematician and cannot solve Exercise 1.
Therefore, Exercise 1 cannot be solved.

l. Anyone who can solve Exercise 1 is a mathematician.
Ralph cannot solve Exercise 1.
Therefore, Ralph is not a mathematician.

m. Some boys like Ralph.
Dogs always like boys.
Ralph is a dog.
Therefore, there is something which both likes and is liked by Ralph.

n. Caesar ruled Rome.
Caesar is dead.
Therefore, a dead person ruled Rome.

o. Ralph is purple.
Purple is a color.
Therefore, Ralph is a color.

5. In formalizing "Marilyn Monroe is an actress", should we consider the ending "-ess" to be part of the form of the proposition and adopt a formalized assumption "$\forall x$ (— is an actress) $(x) \to$ (— is a woman) (x)", or would that be part of an analysis of the word "actress"? Compare "Victoria is a queen".

6. Formulate conventions for the formalization of propositions that contain the pronouns "he", "she", and "who" and then formalize the following.
   a. He who is honest is not a politician.
   b. If someone does not understand Exercise 4, then she will not solve it.
   c. If someone likes turnips, then he'll like rutabagas.
   d. He who dreams sleeps.
   Therefore, he who sleeps dreams.

7. Compare these two conceptions of formalizing.
   a. Formalizing is how we reveal the logical form of an abstract proposition that is represented by an ordinary language sentence.
   [See the *Aside* sections at the end of Chapter 1 (p. 5), Chapter 10.B (p. 73), and Chapter 11.E (p. 86).]
   b. Formalizing is how we reveal the semantic assumptions of ordinary English propositions.

15 Identity

    A. Identity 138
    B. Classical Predicate Logic with Equality 139
    Exercises 142
    C. Implicit Identity vs. Explicit Identity 142
    Exercises 143

A. Identity

Two things are never the same. Or are they?

When I assert "Marilyn Monroe was Norma Jean Baker", I know that the condition for this to be true is that the two names refer to one and the same person, not two.

But suppose I say that I have the same car as you. Do you think I'm revealing an unsuspected joint ownership? No, I mean that I have the same kind of car: with respect to the criteria of year, brand, model, and perhaps color, my car does not differ from yours. That is, if we ignore all but these properties of our cars, your car is identical to mine.

Or I might say without fear of contradiction that your dog is the same as mine. In this case I mean the same breed. The relevant criteria for being the same would probably be body shape, coloring, shape of ears, shape of jaw, and height, but not age or what name it responds to.

On the other hand, suppose I'm talking about my dog and you're talking about a dog you saw last week, and we wonder whether "they" are the same. What do we do? I might start describing mine: it's 45 cm. tall, it's black and white, it's a border collie, it's female, ... If the dog you saw fits the same description for a sufficiently large number of properties, we conclude that it's the same dog. We might be mistaken because we haven't listed enough properties: perhaps you didn't notice enough about the dog you saw, or there's no way to test whether the dog you saw responds to the name "Juney" since you didn't call it. But we nonetheless believe that if they were different we could distinguish them.

But they are different, even if both are "my" dog: she has shed hair since last week, the cut on her paw has healed . . . . Ridiculous, you say, that doesn't matter. They are the same. We know what it is to be a dog, a single, individual dog distinct from all others. The criteria are implicit, granted, but they certainly include the properties we first checked and exclude these latter.

It's not just dogs: stones, chairs, the sun, people. All change from moment to moment. Anything of this world partakes of change and becoming: you can't step into the same river twice. Only in Plato's heaven is there permanence and immutability. There $7 + 3 = 10$ and perhaps in some outskirts of that heaven there is one form that answers to the name "Marilyn Monroe" and also to the name "Norma Jean Baker", though in the world of becoming she changed.

Still, we have an idea of what it is to be one stone, one chair, one sun, one person, an identity of an object that ignores certain differences, an identification according to some implicit criteria. This idea of identity for stones, for chairs, for people must be taken into account when we reason about things. Indeed, we don't have an idea of what we mean by "stone" unless we know what "the same stone", "a different stone", and "two stones" mean. We can't evaluate the truth-value of "$\forall x\, (x$ is a dog $\wedge\ x$ barks)" unless we understand what it means to assign the same object as reference to the two appearances of x in it.

Assertions about identity carry with them implicit criteria of what properties of the objects (that is, which predicates) can be ignored and which matter. Rarely do we make those criteria explicit beyond saying what kind of thing we are considering: "— is the same *dog* as —", "— is the same *stone* as —", "— is the same *person* as —".

B. Classical Predicate Logic with Equality

Neither the notion of thing nor the notion of reference can be studied within a semi-formal language. The notion of thing is meant to be formalized by predicate logic as a whole; reference connects language to the world. In any case, "refers to" is excluded from realizations by the Self-Reference Exclusion Principle. But the predicate "— is the same as —" does not (appear to) create problems with self-reference.

Since identity is fundamental to predicate logic, we want to give an unambiguous explanation of how to interpret "— is the same as —", just as we tried to make explicit how we would interpret "and" when we replaced that word with the formal symbol $\wedge$. Let's replace the predicate "— is the same as —" with a new formal symbol "$\equiv$" called the *equality predicate*, and we'll write "$t \equiv u$" rather than "$(- \equiv -)\,(t, u)$" for terms t and u. We can continue to use the symbol "=" in informal discussions as an abbreviation of "equals".

The phrases "is the same as", "is identical to", and "equals" are essentially synonymous in our ordinary speech. We'll abstract from what's common to them all and grammatical variations of them such as "— and — are the same". So we'll formalize:

> 7 + 3 equals 12.
> The morning star is identical to the evening star.
> Howie and Ralph are the same.

as

> $7 + 3 \equiv 12$
> the morning star $\equiv$ the evening star
> Howie $\equiv$ Ralph

Sometimes the verb "is" is meant as "is the same as". It seems apt to formalize:

as
> Marilyn Monroe was Norma Jean Baker.
> 7 · 2 is 16.

> Marilyn Monroe ≡ Norma Jean Baker
> 7 · 2 ≡ 16

In common speech, "— is different from —" is the antonym of "— is the same as —". So we'll use "¬ (— ≡ —)" to formalize that as well as "— is distinct from —", and variations on these. We can abbreviate "¬ ($t \equiv u$)" as "$t \not\equiv u$".

When we choose a realization, we specify a universe. With that specification we have an implicit idea of identity, implicit criteria for distinguishing any one object of the universe from any other. So we'll interpret "≡" as "— is the same γ as —", where γ is a noun or noun phrase for the objects of the universe.

Example 1: The universe is all objects on my desk.
Analysis We interpret "≡" as "— is the same object on my desk as —"

Example 2: The universe is all real numbers.
Analysis We interpret "≡" as "— is the same real number as —".

Example 3: The universe consists of Ralph, all canaries, and the Eiffel Tower.
Analysis We interpret "≡" as "— is the same Ralph, or the same canary, or the same Eiffel Tower as —".

Example 4: The universe is all men and women.
Analysis We interpret "≡" as "is the same man as or is the same woman as". We do not interpret it as "is the same man or woman as".

Is the equality predicate part of the logical vocabulary or is it categorematic? The question is not just a matter of terminology.

> What we deem to be valid depends on what we may realize as categorematic.

> The criteria of formalization require that categorematic words be retained, whereas syncategorematic words or their formal equivalents can be used in the formalization of any proposition.

Though we vary the interpretation of "≡" from model to model, it is always the implicit identity of the universe. We don't allow it to be interpreted as "is the sister of" or "is the same height as". The interpretation is fixed by the choice of universe, not by any choice for valuations in the model. So we'll view "≡" as syncategorematic, and hence it will be part of the formal language.

Now we can define the formal language and semantics for incorporating the equality predicate into classical predicate logic.

The language of predicate logic with equality The equality predicate ≡ is part of the formal language, the definition of which now includes a new clause:

$u \equiv v$ is an *atomic wff* for any terms u and v.

Interpretation of the equality predicate In any model, for any assignment of references σ and terms u and v,

$\vDash_\sigma u \equiv v$ iff $\sigma(u)$ is the same object of the universe as $\sigma(v)$.

Classical predicate logic with equality is this language and collection of models along with the associated notions of tautology and semantic consequence defined in the usual way.

To axiomatize classical predicate logic we look to two tautologies. First,

Identity $\forall x_1 (x_1 \equiv x_1)$

Note that this is a proposition with no categorematic part.

Principle of Substitution of Equals
$\forall \ldots \forall x \forall y (x \equiv y \rightarrow (A(x) \rightarrow A(y/x)))$
where A is atomic and y replaces some but not necessarily all occurrences of x in A for which it is free

By the consistency of predications, for any atomic predicate P and any terms $t_1, \ldots t_n$, if $\vDash_\sigma x \equiv y$, then $\vDash_\sigma P(t_1, \ldots t_k, x, t_{k+1}, \ldots t_n)$ iff $\sigma \vDash P(t_1, \ldots t_k, y, t_{k+1}, \ldots t_n)$. The inductive definition of truth in a model then ensures that this holds for every wff A: if $\sigma \vDash u \equiv v$, then $\sigma \vDash A(u)$ iff $\sigma \vDash A(v/u)$, where v replaces some but not necessarily all occurrences of u in A for which it is free.

Example 5: Marilyn Monroe ≡ Norma Jean Baker Therefore,
(— *was an actress*) (*Marilyn Monroe*) → (— *was an actress*) (*Norma Jean Baker*)
Analysis This is valid.

Example 6 $\forall x \forall y [x \equiv y \rightarrow ((- \text{ is a dog}) (x) \rightarrow (- \text{ is a dog}) (y))]$
Analysis This is a tautology.

Axioms for equality

5. $\forall x (x \equiv x)$ *identity*

6. $\forall \ldots \forall x \forall y (x \equiv y \rightarrow (A(x) \rightarrow A(y/x)))$ *substitution of equals*
   where A is atomic and y replaces some but not
   necessarily all occurrences of x in A for which it is free

---

In Appendix 4 you can find a proof that this axiom system is complete: $\Gamma \vDash A$ iff $\Gamma \vdash A$.

Exercises
1. a. List four predicates of English that you consider to be (sufficiently) synonymous with "is the same as" to be formalized as ≡.
   b. List four predicates of English that you believe are (sufficiently) synonymous with "is not the same as" to be formalized as ≢.
2. Give criteria of identity for a universe consisting of:
   a. All objects in your bedroom.
   b. All parts of your pen.
   c. All the pigs in Denmark.
   d. All integers.
   e. All irrational numbers between 0 and 1.
   f. Atoms.
3. Explain why we classify ≡ as syncategorematic.

C. Implicit Identity vs. Explicit Identity
We started with some idea of identity, set out what we consider its most salient properties, and got a theory. That's just what we do when we give a formal theory of "— is parallel to —" in geometry. As in geometry, there might then by different ways to interpret the formal notion in a model that satisfy the axioms. Let' see what happens if we do that with "≡".

Consider this realization and universe:

L(¬, →, ∧, ∨, ∀, ∃ ; — is a dog, — is a cat, — is a sheep,
— is bigger than 3 cm tall)

universe all domestic animals on my ranch, Dogshine

Suppose we interpret the predicates as we normally do. If we add ≡ to the realization, we could interpret it any way that satisfies the axioms. So we could have that the assignment of my dog Birta to x and my dog Chocolate to y satisfies "$x \equiv y$", since no atomic predicate distinguishes them. We could, though we need not, say that the assignment of my dog Birta to x and my dog Bidú to y also satisfies "$x \equiv y$", so that Birta, Chocolate, and Bidú are all counted as the same. We could say that any two, or three, or all seven of my chickens are to be counted as the same, for they all satisfy the same atomic predicates. The maximal relation satisfying the axioms in the sense of identifying the most objects is:

Explicit identity
$\vDash_\sigma x \equiv y$ iff $\sigma(x)$ and $\sigma(y)$ satisfy the same atomic predicates

That is, $\vDash_\sigma x \equiv y$ iff

> given any wff A(x) in which x appears free, for any assignment of references τ, such that τ(x) = σ(y) and τ agrees with σ on all other variables, $\vDash_\sigma (A(x)) = T$ iff $\vDash_\tau (A(x)) = T$.

To adopt this would be to take as one rather than two whatever we cannot distinguish with the predicates we have in our model. This would make explicit what we mean by identity, rather than relying on the implicit identity of the universe.

But we still need the implicit identity of the universe in order to give assignments of reference. And allowing this maximal interpretation of $\equiv$ or any other that satisfies the axioms for equality would mean that $\equiv$ is no longer syncategorematic. Indeed, it would be the most categorematic of all predicates because to establish it we have to already have the interpretation of all the other categorematic parts of the semi-formal language. We could have one realization and universe and two models for that such that if even one of the predicates is interpreted differently in one rather than the other, then the maximal interpretation of $\equiv$ would be interpreted differently, too. For instance, if we interpret "— is a cat" to mean what we would normally use "— is black and white" for, then the assignment of my dog Birta to x and my dog Bidú to y would no longer satisfy "$x \equiv y$", for Birta is brown and Bidú is black and white. And if $\equiv$ is not syncategorematic, we could not use the following to say that there are at least two dogs:

$$\exists x \exists y ((x \not\equiv y) \wedge (\text{— is a dog}) (x) \wedge (\text{— is a dog}) (y)$$

In the first model described above, though there are three dogs, they are all identified by the maximal interpretation of $\equiv$, which is just to say that "for the purposes at hand" there is only one dog in the universe of the model.

These reflections reinforce our decision to interpret $\equiv$ as the implicit identity of the universe.

Exercises
1. An *equivalence relation* on a collection is any relation $\simeq$ that satisfies:

$a \simeq a$	*identity*
If $a \simeq b$, then $b \simeq a$.	*symmetry*
If $a \simeq b$ and $b \simeq c$, then $a \simeq c$.	*transitivity*

    If $\simeq$ is an equivalence relation on a collection U, define:

    $[a] = \{b : b \simeq a\}$ *the equivalence class of a*

    a. Show that if $\simeq$ is an equivalence relation on a collection U, then for each a in U, a is in one and only one equivalence class of $\simeq$.
    b. Show that if U is divided into non-empty subsets so that each a in U is in one and only one of those subsets, then the following is an equivalence relation:
        $a \simeq b$ iff a and b are in the same subset
    c. Show that in a model the explicit identity interpretation of $\equiv$ is an equivalence relation on the universe.
    d. Give an example of an equivalence relation on a universe of a model that does not satisfy the extensionality condition.

Key Words identity principle of substitution of equals
implicit identity explicit identity
equality predicate

16 Formalizing with the Equality Predicate

A. Formalizing Other Quantifiers
   There are at least n 144
   There are at most n 145
   There are at exactly n 146
   "no" and "nothing" 146
   Quantifications we can't formalize 147
B. Examples of Formalizing 148
Exercises 153

A. Formalizing Other Quantifiers

We've chosen as basic just two ways of specifying the number of things that satisfy a condition: everything or at least one. In this section we'll look at other ways of saying how many.

There are at least n

Example 1: *There are at least two dogs.*
Analysis We can formalize this with:

(*) $\exists x \exists y \, [x \not\equiv y \,\wedge\, (-\text{ is a dog})(x) \,\wedge\, (-\text{ is a dog})(y)]$

This is true in a model iff there are at least two objects in the universe that satisfy "— is a dog". Note, though, that to view this as a good formalization according to our criteria, we must view "two" in the proposition as part of a quantifier, syncategorematic, rather than as a name.

Example 2: *There are at least three dogs.*
Analysis We can formalize this with:

(*) $\exists x \exists y \exists z \, [x \not\equiv y \,\wedge\, x \not\equiv z \,\wedge\, y \not\equiv z \,\wedge\, (-\text{ is a dog})(x) \,\wedge\,$
        $(-\text{ is a dog})(y) \,\wedge\, (-\text{ is a dog})(y)]$

This is true in a model iff there are at three objects in the universe that satisfy "— is a dog".

    Since we have plenty of variables, we can formalize "There are at least n" for any counting number n. We first define a formula that is true iff the distinct variables $y_1, \ldots, y_n$ are assigned distinct references. Intuitively, for $n \geq 2$,

$E_n(y_1, \ldots, y_n)$ is:
$y_1 \not\equiv y_2 \,\wedge\, y_1 \not\equiv y_3 \,\wedge\, y_2 \not\equiv y_3 \,\wedge\, y_1 \not\equiv y_4 \,\wedge\, \ldots \,\wedge\, y_{n-1} \not\equiv y_n$

An inductive definition of E_n for $n \geq 2$

$E_2(y_1, y_2) \equiv_{\text{Def}} y_1 \not\equiv y_2$

$E^1_{n+1}(y_1, \ldots, y_{n+1}) \equiv_{\text{Def}} E_n(y_1, \ldots, y_n) \wedge (y_1 \not\equiv y_{n+1})$

$E^{m+1}_{n+1}(y_1, \ldots, y_{n+1}) \equiv_{\text{Def}} E^m_{n+1}(y_1, \ldots, y_{n+1}) \wedge (y_{m+1} \not\equiv y_{n+1})$

$E_{n+1}(y_1, \ldots, y_{n+1}) \equiv_{\text{Def}} E^n_{n+1}(y_1, \ldots, y_{n+1})$

For any model and assignment of references σ:

$v_\sigma \vDash E_n(y_1, \ldots, y_n)$ iff σ assigns distinct references to $y_1, \ldots, y_n$

For any formula $A(x)$ in which x is the only free variable, we can now assert that there are at least n things that satisfy $A(x)$. Let $y_1, \ldots, y_n$ be the first n variables that do not appear in A in order of increasing indices.

Convention: Formalizing "there are at least n"

$\exists_{\geq 1} x\, A(x) \equiv_{Def} \exists x\, A(x)$

$\exists_{\geq n} x\, A(x) \equiv_{Def}$
$\qquad \exists y_1 \ldots \exists y_n\, (E_n(y_1, \ldots, y_n) \wedge (A(y_1/x) \wedge \ldots \wedge A(y_n/x)))$

Then we have:

$\exists_{\geq n} x\, A(x)$ is true in a model iff
iff there are at least n things in the universe that satisfy $A(x)$

Example 3: *There are at least 47,813,201 stars.*
Analysis We can formalize this:

(*) $\quad \exists_{\geq 47,813,201} x\, (-\text{ is a star})(x)$.

Example 4: *There are infinitely many stars.*
Analysis We cannot formalize this. At best we can point to the collection:

$\exists_{\geq 1} x\, (-\text{ is a star})(x)$
$\exists_{\geq 2} x\, (-\text{ is a star})(x)$
$\exists_{\geq 3} x\, (-\text{ is a star})(x) \quad \ldots$

There are at most n

Example 5: *There are at most two nice cats.*
Analysis We can take this as an assertion that there are not more than two nice cats. That is, there aren't even three nice cats.

(*) $\quad \neg \exists_{\geq 3} x\, [\,(-\text{ is a cat})(x) \wedge (-\text{ is nice})(x)\,]$

This is true in a model iff there are at most two objects in the universe that satisfy both "— is a cat" and "— is nice". Moreover, the formalization uses the same categorematic words as the original, viewing the quantification "There are at most two" as syncategorematic.

Convention: Formalizing "there are at most n"

For every wff A with just one variable x free, and for every $n \geq 1$, define:

$\exists_{\leq n} x\, A(x) \equiv_{Def} \neg \exists_{\geq n+1} x\, A(x)$

Then we have:

$\exists_{\leq n} x\, A(x)$ is true in a model
$\qquad$ iff there are at most n things in the universe that satisfy $A(x)$

There are exactly n

Example 6: *There are exactly 47 dogs that do not bark.*
Analysis We can say that there are at least 47 dogs that don't bark. We can say there are at most 47 dogs that don't bark. So we can say there are exactly 47 dogs that don't bark:
(*) $\exists_{\geq 47} x\, ((-\text{ is a dog})\,(x) \land \neg(-\text{barks})\,(x)) \land$
     $\exists_{\leq 47} x\, (-\text{ is a dog})\,(x) \land \neg(-\text{barks})\,(x))$

Convention: *Formalizing "there exactly n"*
For any $n \geq 1$ and any wff A in which x is the only variable free,
$$\exists!_n x\, A(x) \equiv_{\text{Def}} \exists_{\geq n} x\, A(x) \land \exists_{\leq n} x\, A(x)$$

Then we have:

$\exists!_n x\, A(x)$ is true in a model
    iff there are exactly n things in the universe that satisfy $A(x)$

It's common to write "$\exists! x$" for "$\exists!_1 x$", reading that as "There is a unique x such that". Writing out the abbreviation in full, we have:

$\exists! x\, A(x) \equiv_{\text{Def}} \exists x\, A(x) \land \neg \exists y\, \exists z\, (y \not\equiv z \land (A(y) \land A(z)))$

where y and z are the variables with least index that do not appear in $A(x)$, with y having index less than z.

Similarly, we can specify how many things are in the universe,

Convention: *Formalizing "there are n things"*
$\exists x_1 \ldots \exists x_n\, E_n(x_1, \ldots, x_n) \land$
$\neg \exists x_1 \ldots \exists x_n\, \exists x_{n+1}\, E_{n+1}(x_1, \ldots, x_n, x_{n+1})$

This is true in a model iff there are exactly n things in the universe

Example 7: *There are at least 7 but fewer than 42 wombats.*
Analysis We can formalize this:
(*) $\exists_{\geq 7} x\, (-\text{ is a wombat})\,(x) \land \exists_{\leq 41} x\, (-\text{ is a wombat})\,(x)$

Note that this requires us to view "there are 47", "at most 12", and "fewer than 42" as syncategorematic rather than using "47", "12", and "42" as names.

"no" and "nothing"

Example 8: *No dog meows.*
Analysis We could take the example as equivalent to "Every dog doesn't meow" and formalize it as in Example 14 of Chapter 14 (p. 113):

$\forall x\, [\, (-\text{ is a dog})\,(x) \rightarrow \neg(-\text{meows})\,(x)\,]$

This is equivalent to:

(*) ¬∃x [(— is a dog) (x) ∧ (— meows) (x)]

But (*) tracks the grammar of the original better, understanding "no" as "no thing" and the example as "No thing is a dog that meows", which is in accord with the convention on negations and respecting grammar.

Example 9: *Nothing both barks and meows.*
Analysis Here the reading of "no" is explicit, and we can formalize the example as:

(*) ¬∃x ((— barks) (x) ∧ (— meows) (x))

Some earlier philosophy was infected with the confusion that a grammatical subject must refer to something, with metaphysicians struggling to say exactly what nothing is, as you can read in a bravura display in the entry "Nothing" by P. L. Heath in *The Encyclopedia of Philosophy*, 1967. Here the puzzle evaporates: such negative subjects are replaced by ¬∃x.

Example 10: *No dog is a parent of a cat.*
Analysis We can formalize "Some dog is a parent of a cat" in accord with our conventions:

∃x [(— is a dog) (x) ∧ ∃y ((— is a parent of —) (x, y) ∧ (— is a cat) (y)]

Then we can formalize the example by placing the negation at the start in accord with the convention on negations:

(*) ¬∃x [(— is a dog) (x) ∧ ∃y ((— is a parent of —) (x, y) ∧ (— is a cat) (y)]

> *Convention*: *Formalizing "no" and "nothing"*
> We formalize "no" or "nothing" used as a quantifier as ¬∃x.
>   No γ is δ.   ¬∃x (γ(x) ∧ δ(x))
>   No γ is R to δ.   ¬∃x [γ(x) ∧ ∃y (R(x, y) ∧ δ(y))]

Quantifications we can't formalize

Example 11 Almost all dogs bark.
Analysis By "almost" we don't mean "all but seven", or "90%", or any other precise quantifier. Quantifications such as "almost all", "many", "most ", "all but a few" are usually dismissed as being too imprecise to incorporate into formal logic.
　　Nonetheless, we can reason well with sentences such as this, as you can see in my *Critical Thinking*, for such reasoning exhibits regularities based on semantic assumptions. Our decision not to incorporate such quantifications is not a principled one, only a desire to avoid problems that don't seem fundamental enough to worry about at this stage.

Example 12 2/3 of all cats have fleas.
Analysis At the opposite extreme from imprecise quantifications are the exact ways of saying how many, like this example and "86% of all teenage girls have smoked

marijuana". Such precision is certainly welcome in reasoning. But if we use a quantifier "40% of all", why shouldn't we include "40.03% of all"? There would be no end but to include a great deal of mathematics into what should be the fundamentals of all applications of logic. So we demur at this stage.

We have seen how to say that exactly, or at least, or at most how many things satisfy an open sentence, or that all things do, or that no things do. In classical predicate logic we have only these quantifying expressions plus complications of them using the connectives and grammar of predicate logic; you can a proof of that in my *Classical Mathematical Logic*.

B. Examples of Formalizing

Example 13: Marilyn Monroe married Arthur Miller.
    Marilyn Monroe was Norma Jean Baker.
    Therefore, *Norma Jean Baker married Arthur Miller*.

Analysis The formalization is straightforward:

(*)   (— married —) (Marilyn Monroe, Arthur Miller)
    Marilyn Monroe ≡ Norma Jean Baker
    Therefore, (— married —) (Norma Jean Baker, Arthur Miller)

This is valid. Is the informal inference valid? We have to argue that it is correct to treat "— married —" as extensional and to treat reference as atemporal in order to ignore the tense of the premises.

Example 14:   Marilyn Monroe is a movie star.
    Marilyn Monroe is Norma Jean Baker.
    Therefore, *Norma Jean Baker is a movie star*.

Analysis We formalize:

(*)   (— is a movie star —) (Marilyn Monroe)
    Marilyn Monroe ≡ Norma Jean Baker
    Therefore, (— is a movie star —) (Norma Jean Baker)

The analysis is the same as in the last example, only here note the two different treatments of "is".

Example 15: *Everything is bigger than something*.

Analysis Without a context that determines otherwise, I understand the example to mean that given anything, there's something it's bigger than. We can then formalize it taking the quantifiers in the order in which they appear in the original.

(*) $\forall x \, \exists y \, (-$ is bigger than $-) \, (x, y)$

Example 16: *Something is bigger than everything*.

Analysis This isn't meant to be false just because nothing is bigger than itself. The comparison is to other things.

(*) $\exists x \, \forall y \, [x \not\equiv y \rightarrow (-$ is bigger than $-) \, (x, y) \,]$

Again, this follows the original for the order of the quantifiers.

Example 17: Nothing is bigger than everything.
Analysis According to how we've chosen to understand "nothing", this example is the negation of the previous one:

(*) ¬ ∃x ∀y [x ≠ y → (— is bigger than —) (x, y)]

Example 18: Nothing is friendlier than Birta.
Analysis The formalization follows our convention on formalizing "nothing", except that here because of the comparative we have to add a clause excluding Birta from the range of the quantifier.

(*) ¬ ∃x (x ≠ Birta ∧ (— is friendlier than —) (x, Birta)

Example 19: Nothing is friendlier than a dog.
Analysis What do we mean by this? A lot of people believe it's true, but they don't believe that nothing is friendlier than the junk yard dog that barks ferociously at everybody that passes the fence. So we can't read this as "Nothing is friendlier than any dog". But then how can we understand it? I think the example is a vague way of saying that in general dogs are friendlier than other animals. We have no way to formalize "in general". Even incorporating mathematics into our system wouldn't help, for it's not a statistical generalization.

Example 20: Nothing is as dumb as a sheep.
Analysis Here I think that this can be understood as "Nothing is as dumb as any sheep". So we can formalize it as:

(*) ∀x [¬ (— is a sheep) (x) →
         (∀y (— is a sheep) (y) → ¬ (— is as dumb as —) (x, y))]

Example 21: Something is better than nothing.
Analysis We can't eliminate "nothing" as a noun here in favor of quantifiers: "∃x ¬ ∃y (x ≠ y ∧ (— is better than —) (x, y)" is simply wrong. The example is not formalizable.

But can't we rewrite the example to make it clear enough to formalize? Perhaps it's meant as "Something is better than having nothing" or perhaps "Having something is better than having nothing". I think this bit of folk wisdom is meaningless.

Example 22: No dog meows.
   Juney is a dog.
   Therefore, *Juney does not meow.*
Analysis The informal inference is valid, and so is its formalization:

(*) ¬ ∃x [(— is a dog) (x) ∧ (— meows) (x)]
     (— is a dog) (Juney)
     Therefore, ¬ (— meows) (Juney)

An equivalent formalization of the first premise is not good:

   ∀x [(— is a dog) (x) → ¬ (— meows) (x)]

150   An Introduction to Formal Logic

The word "no" in "No dog meows" applies to "dogs": it is a kind of quantifier, not a negation of "meows".

Example 23:   *There is no honest person who is a politician.*
Analysis The formalization is straighforward according to our conventions:
(*)   ¬∃x [(— is a person) (x) ∧ (— is honest) (x) ∧ (— is a politician) (x)]

Example 24:   *Every honest person is not a politician.*
Analysis Our conventions lead us to formalize this as:
(*)   ∀x [((— is a person) (x) ∧ (— is honest) (x)) → ¬(— is a politician) (x)]

This and the last example are informally equivalent, and their formalizations are equivalent, too. So why don't we formalize them the same? That would be to do analysis before formalization. We follow the grammar of what we are given unless we have no other way to proceed in accord with how we understand the sentences.

Example 25:   *There is no tallest person.*
Analysis Again we take "there is no" to mean "Not: there is":
(*)   ¬∃x [(— is a person) (x) ∧
         ∀y (((— is a person) (y) ∧ y ≢ x)→ (— is taller than —) (x, y))]
Superlatives come from comparatives, in this case "is taller than", and as with those, the comparison must be made to all objects other than the object claimed to be the superlative.

Example 26:   *Only Ralph barks.*
Analysis I understand the example to be equivalent to:
   Ralph barks and nothing else barks.
The "else" means that the assertion is about things that aren't Ralph. So we can rewrite this as:
   Ralph barks and nothing that is not Ralph barks.
We've agreed to formalize "nothing" as ¬∃x, and since "Ralph" is a name, we take the "is" in "is Ralph" as ≡.
(*)   (— barks) (Ralph) ∧ ¬∃x [(x ≢ Ralph) ∧ (— barks) (x)]

Example 27 *Only dogs bark.*
Analysis We proceed in parity with the last example and get:
   Dogs bark and nothing that is not a dog barks.
The problem, then, is whether to read "Dogs" as "All dogs" or "Some dogs". Compare:
   Only rich people have a Rolls-Royce.
   Only women bear children.
We don't mean "only" in these as "all": we're not asserting that all rich people own a Rolls-Royce but rather that some do and no one else does. So we formalize the example:

(a) $\exists x [(-\text{ is a dog})(x) \land (-\text{ barks})(x)] \land$
    $\neg \exists y [\neg (-\text{ is a dog})(y) \land (-\text{ barks})(y)]$

As a general rule, we'll include an existential assumption in formalizing "only" unless context demands otherwise. Those who intend "only" to be taken in the sense of "all" shall have to make that explicit, rewriting, for example "Only cats are nasty" as "All cats are nasty and nothing that is not a cat is nasty". But consider:

   Only a supervisor can approve a pay raise for Maria.

The position of supervisor has not been filled for a month. So no one can approve a pay raise for Maria. Does that mean the example is false? I think not. This suggests that "only" does not include "at least one". If so, we should formalize the example as:

(b) $\neg \exists y (\neg (-\text{ is a dog})(y) \land (-\text{ barks})(y))$

Example 28: *Only dogs hate cats.*
Analysis We have to decide whether "cats" means "all cats" or "some cats". Is the example false because Fido hates all cats except Puff? I think not. So let's take "cats" to mean "some cats". Then, as in the last example we have the choice of whether an existential assumption should be included for "dogs":

(a) $\exists x ((-\text{ is a dog})(x) \land \exists y (-\text{ is a cat})(y) \land (-\text{ hates }-)(x, y)) \land$
    $\neg \exists z [\neg (-\text{ is a dog})(z) \land \exists y (-\text{ is a cat})(y) \land (-\text{ hates }-)(x, y)]$

(b) $\neg \exists z [\neg (-\text{ is a dog})(z) \land \exists y (-\text{ is a cat})(y) \land (-\text{ hates }-)(x, y)]$

There is no principled choice between (a) and (b) in this example or in the last example. As with "all", we can use a formalization of "only" in which we do or do not consider an existential assumption as part. Minimizing existential assumptions unless context demands, in parity with "all", suggests that we take (b) in both of these examples as the formalization. But there is no consensus on this.

Example 29: *Not only dogs bark.*
Analysis I don't take this example as the negation of Example 28 because the following deduction is valid:

   Not only dogs bark.
   Therefore, dogs bark.

Here "not" is meant as part of the compound quantifier "not only" rather than as a negation of the whole. So we rewrite this example as:

   Dogs bark and not: nothing that is not a dog barks.

(*) $\forall x [(-\text{ is a dog})(x) \rightarrow (-\text{ barks})(x)] \land$
    $\neg \forall y [\neg (-\text{ is a dog})(y) \rightarrow \neg (-\text{ barks})(y))]$

Example 30: *Ralph is the father of Birta.*
Analysis The use of "the" indicates uniqueness: If Ralph is the father of Birta, then nothing else is a father of Birta. We have to respect that with the formalization.

(*) $(-\text{ is a father of }-)(\text{Ralph}, \text{Birta}) \land$
    $\neg \exists x [(-\text{ is a father of }-)(x, \text{Birta}) \land (x \equiv \text{Ralph})]$

Example 31: *Everyone has two parents.*
Analysis It's not right to read "has" as "possesses" as in Example 50 of Chapter 14 (p. 128). Rather, to say that you have a parent is just to say that there is a parent of you. The formalization is then straightforward using our convention of reading "every one" as "every person" and a meaning axiom to relate "is a parent of" to "is a parent".

(*) $\forall x\, [\, (-\text{ is a person})\, (x) \to \exists_{\geq 2} y\, (-\text{ is a parent of }-))\, (y, x)\,]$

relative to $\forall x\, \forall y\, [\, (-\text{ is a parent of }-))\, (y, x) \to (-\text{ is a parent}))\, (y)\,]$

In this formalization, I've understood "two" as "at least two", for the sentence does not assert that everyone has exactly two parents; adding that would be analysis before formalization. I see no general rule governing when to interpret a numerical quantification as "at least" or "exactly".

Example 32 *Some cats are feral.*
Analysis This was Example 4 of Chapter 14 (p. 109). We can now formalize the use of "some" with a plural common noun as meaning "more than one":

(*) $\exists_{\geq 2} x\, [\, (-\text{ is a cat})\, (x) \wedge (-\text{ is feral})\, (x)\,]$

Example 33 *Some dogs like some cats.*
Analysis The formalization follows the conventions for converting nouns into predicates.

(*) $\exists x\, \exists y\, [\, (-\text{ is a dog})\, (x) \wedge (-\text{ is a cat})\, (y) \wedge (-\text{ likes }-)\, (x, y)\,]$

If you think that the plural forms indicate more than one, you'll have to decide how many dogs must like how many cats for the proposition to be true. One reading is that there are at least two dogs each of which likes at least two cats:

$\exists_{\geq 2} x\, \exists_{\geq 2} y\, [\, (-\text{ is a dog})\, (x) \wedge (-\text{ is a cat})\, (y) \wedge (-\text{ likes }-)\, (x, y)\,]$

Example 34 *Ralph is a dog and Juney is a dog.*
   Therefore, *there are two dogs.*
Analysis I take "there are two" in the example to mean "there are at least two".

(*) $(-\text{ is a dog})\, (\text{Ralph}) \wedge (-\text{ is a dog})\, (\text{Juney})$
   Therefore, $\exists_{\geq 2} x\, (-\text{ is a dog})\, (x)$

The deduction is not valid, and writing out the conclusion in full will show why:

$\exists y\, \exists z\, [\, (y \neq z) \wedge (-\text{ is a dog})\, (y) \wedge (-\text{ is a dog})\, (z)\,]$

For the inference to be valid we need an additional premise: $(\text{Ralph} \neq \text{Juney})$.

Example 35: *All but 2 bears in the London Zoo are brown.*
   There are 47 bears in the London Zoo.
   Therefore, there are 45 brown bears in the London Zoo.
Analysis I understand "All but 2 bears are brown" as "2 bears are not brown and all the rest are brown". That is, there are exactly 2 bears that are not brown.

(*) $\exists!_2 x\, [\, (-\text{ is a bear})\, (x) \wedge (-\text{ is in the London Zoo})\, (x) \wedge \neg (-\text{ is brown})\, (x)\,]$

(I haven't taken "the London Zoo" as a name in order to simplify the discussion.) The inference is valid in classical propositional logic.

By viewing "2", "47", and "45", as parts of syncategorematic expressions, a problem in arithmetic can be converted to an investigation of logical validity. We convert counting things into counting variables.

Example 36: *Through any two points there is exactly one line.*
Analysis We can formalize this as:

(*) $\forall x \, \forall y \, [\,((- \text{ is a point})(x) \land (- \text{ is a point})(y) \land x \not\equiv y) \rightarrow$
    $\exists !z \, ((- \text{ is a line})(z) \land (- \text{ is through } -)(z,x) \land (- \text{ is through } -)(z,y))\,]$

I understand "any two" as "any two distinct things". Mathematicians sometimes say "any two" when they mean any things referred to by two variables, as in "Any two points are identical or determine a line". Sometimes you can find them really struggling with language, saying "For any two not necessarily distinct . . .".

Example 37 *There are 2 prime numbers that differ by 2.*
Analysis We can formalize this as:

(*) $\exists x \, \exists y \, [\,(- \text{ is a number})(x) \land (- \text{ is a number})(x) \land x \not\equiv y$
    $\land (- \text{ is a prime})(x) \land (- \text{ is a prime})(y) \land (- \text{ differs from } - \text{ by } -)(x, y, 2)\,]$

Here one occurrence of "2" is formalized as syncategorematic and one occurrence as categorematic. To formalize "differs from" as subtraction, we'd need to have a way to formalize talk of functions, which you can see in my *The Internal Structure of Predicates and Names*.

Example 38 *There are more dogs than cats.*
Analysis "There are more" is a precise quantification, but we have no tools for formalizing it in predicate logic, for it is comparing sizes of collections of things.

Exercises
1. Formalize the following or explain why no good formalization is possible.
    a. Everybody loves somebody.
    b. Everybody loves somebody other than himself.
    c. Everybody loves somebody sometime.
    d. Everyone knows someone who is famous.
    e. Ralph is Ralph, not a dog.
    f. Some cats are nice.
    g. Nobody likes logicians.
    h. Nobody knows the trouble I have seen.
    i. No one who kicks a cat is a philosopher.
    j. Anyone who kicks a cat is not a philosopher.
    k. Nothing barks unless it is a dog.
    l. Nobody in New Mexico is taller than everyone in Nevada.
    m. There is a tallest person.
    n Ralph is smarter than any dog in Cedar City.

154   An Introduction to Formal Logic

    o. Ralph is the smartest dog in Cedar City.
    p. Only women sew.
    q. Some people love only dogs.
    r. Only some people love dogs.
    s. In China some couples have more than two children.
    t. Four cats are sitting in a tree.
    u. No three people in Cedar City have the same father.
    v. There are at least 33 but fewer than 412 cats that haven't scratched someone.
    w. You may fool all the people some of the time; you can even fool some of the people all the time; but you can't fool all the people all the time. (Abraham Lincoln)
    x. God is that than which nothing greater can be conceived.
    y. The mind and the body are distinct.

2. Formalize the following and say whether the inference is valid or invalid, or explain why no good formalization is possible.

    a. Mark Twain wrote *Huckleberry Finn*.
       Mark Twain is Samuel Clemens.
       Therefore, Samuel Clemens wrote *Huckleberry Finn*.

    b. There is only one President of the U.S.
       George Bush is President of the U.S.
       George McGovern is not George Bush.
       Therefore, George McGovern is not president of the U.S.

    c. Only Ralph and Juney are dogs.
       Both Ralph and Juney bark.
       Therefore, all dogs bark.

    d. Ralph is not Juney.
       Juney is not a puppet
       Therefore, Ralph is a puppet.

    e. All but 4,319 horses do not live in Utah.
       There are 18,317,271 horses.
       Therefore, there are 18,312,956 horses that do not live in Utah.

    f. Every father has at least two sons.
       Any two sons of the same father are brothers.
       Every man is a son.
       Therefore, every man has a brother.

    g. All cats have fleas.
       Juney is a dog.
       No dog is a cat.
       Therefore, Juney does not have fleas.

    h. Nothing is better than an ice cream cone.
       Dick gave Suzy nothing.
       Therefore, Dick gave Suzy something better than an ice cream cone.

3. Suppose we take the collection of my dogs as the universe for a realization. There are three of them: Birta, Chocolate, and Bidú. Suppose also that their names are in the semi-formal language. Then:

∀x (x is a dog)   is true in the model iff
Birta is a dog ∧ Chocolate is a dog ∧ Bidú is a dog

∃x (x is a dachshund)   is true in the model iff
Birta is a dachshund ∨ Chocolate is a dachshund ∨ Bidú is a dachshund

For any finite universe all of whose objects have names in the semi-formal language, we can dispense with quantification: ∀x can be replaced with a conjunction, and ∃x can be replaced with a disjunction. For this reason, some people write ⋀ x for the universal quantifier and ⋁ x for the existential quantifier.

a. With the universe and names described above, give formalizations of the following that do not use quantifiers:

Everything is a mammal.
Nothing is a cat.

Would they be acceptable according to our criteria of formalization?

b. Make up a universe with 4,317 objects, each of which has a name in the semi-formal language and formalize without quantifiers:

Everything is a mammal.
Nothing is a cat.

17   Possibilities

To reason is to make inferences.

> An *inference* is a collection of two or more propositions, one of which is the conclusion and the others the premises, that is intended by the person who sets it out as either showing that the conclusion follows from the premises or investigating whether that is the case.

For an inference to be good, the inference must be valid or strong.

> An inference is *valid* means that there is no way the premises could be true and the conclusion false at the same time.
>
> An inference is *strong* means that there is a way for the premises to be true and the conclusion false, but all such ways are unlikely.

To evaluate an inference, we must consider ways the world could be.

We've seen through examples that to invoke a way the world could be, a possibility, we have no choice but to use a description when we wish to reason together. A description of the world is a collection of propositions: we suppose that this, and that, and this are true. We do not require that we give a complete description of the world, for no one is capable of presenting such a description nor is anyone capable of understanding one if presented. By using collections of propositions to stand in for or as descriptions of possibilities, we need not commit ourselves to a possibility being something real, such as a world in which I am not bald.

But what qualifies some collections of propositions as describing a possibility and others not? What do we mean by saying that a dog giving birth to a donkey is a possibility, but a square circle is not? If you say that a dog giving birth to a donkey is not a possibility, how are we to decide if you are right?

Perhaps we have different ideas about what is possible. You might consider that it is physically impossible for a dog to give birth to a donkey, knowing all we know now about the biology of these animals. I might say that it is possible; we just don't know how yet. Or I might say that it is not possible for a dog to give birth to a donkey by any ethical means, whereas you might say that it would be perfectly acceptable morally to interfere with the biology of dogs and donkeys to bring that about.

There are many different notions of possibility: physical possibility, moral possibility, possibility given what has happened up to this time, ... Regardless of which of these we are employing, we always seem to agree that a description of a way the world could be must be at least consistent. That is, it cannot have or entail a contradiction. It must be *logically possible*.

This seems to be the ground from which we all start in our reasoning. What is possible must be consistent. So there is no way the world could be in which there is a square circle. But there seems to be no contradiction inherent in postulating that a dog could give birth to a donkey: it is logically possible.

It might seem we have made some progress in analyzing valid inferences. But the progress is illusory. What is logically possible is what contains or entails no contradiction. But that requires knowing what it means for a collection of propositions to entail another, which is what we are trying to understand. What is logically possible depends on how we understand valid inferences. But what is a valid inference depends on how we understand possibilities, particularly logical possibilities. We seem to be in a circle with no way out.

One way we could extricate ourselves from the circle is to agree that we will use our informal, intuitive reasoning to determine what is logically possible. But that is to deny the entire project of trying to understand valid inferences, for we would be trying to make explicit a concept that depends on our not making explicit our most fundamental assumptions.

What we can do is investigate parts of our reasoning, picking out just this or that kind of reasoning relative to restricted semantic assumptions that allow for clarity of analysis, calling that "a logic". Then we can have a clearer notion of possibility and of valid inference for that kind of reasoning. As we extend our investigations to allow for more kinds of reasoning, we will have fuller analyses of logical possibilities and valid inferences. But unless we should ever formalize all of reasoning, which seems very unlikely, we shall never have a complete analysis of logical possibilities and valid inferences. We shall have analyzed those at best only relative to the specific assumptions we have made.

Can we find some properties of propositions that might be simpler and in terms of which we could understand possibilities? The most basic is that each proposition is (or is to be considered) true or false.

What further simple or clear properties of claims may be of use? One good candidate is their linguistic structure.

Some consider linguistic structure enough. They set out formal systems of reasoning, logics, solely in terms of linguistic form. They give syntactic forms of propositions that are deemed true for every possibility and syntactic forms of inferences that are deemed valid. These are then taken as the norms of reasoning. This, they say, is a way to extricate ourselves from the possibility-inference circle.

Though it may allow for a way out of the circle, it promises no insight or further clarity about inferences or possibility. It leaves the circle only to embrace intuitive justifications for why we should accept propositions of those particular forms as always true and inferences of those forms as valid. We are back at accepting an informal understanding of valid inference or an informal notion of possibility as the basis of our reasoning. If, then, we look at the structure of propositions, it is to reduce possibilities and valid inferences to simpler semantic notions. This is what we've done in this text.

We started with linguistic forms of propositions built from other propositions using connectives. The only semantic value of propositions we assumed is that each is (or is considered to be) true or false. Assuming that the truth-value of the whole is determined by the truth-values of its parts, and assuming the division of form and content, we developed classical propositional logic.

When formalizing reasoning in classical propositional logic, a possibility is reduced to a way to assign truth-values to atomic propositions, resulting in a model for the logic. To say that an inference is valid relative to classical propositional logic is to say that there is no model in which the premises are (taken to be) true and the conclusion false. That is, there is no way to assign truth-values to the atomic propositions in the semi-formal language that will make the premises true and the conclusion false. We have a reduction of the notion of possibility to simpler notions.

This analysis still invokes our informal notion of possibility, for we talk about any possible assignment of truth-values. But this is a restricted and much simpler notion than the idea of possibility in general. An inference has a finite number of premises plus a conclusion, say n propositions in all. There are 2^n different ways to assign truth-values to those, which we can inspect in a tabular form when n is small enough. When n is very large or where we choose to work with infinitely many premises, as we sometimes do in mathematics, we must rely on the method of specification of the atomic propositions in them to guide us in how we can assign truth-values to them. The reduction of the notion of validity to simpler notions will then be as clear and convincing as our understanding of large numbers or infinitely many premises.

We can expand our analysis to take into consideration further aspects of propositions, such as how we may come to know them, or their subject matter, or how likely we deem them to be true. We do that by making structural decisions about the nature of the additional aspect of propositions we are considering and then factoring those into slightly more complicated truth-tables. For each such analysis, that is, for each such logic, there is a corresponding notion of a model, which formalizes the notion of possibility and, hence, a notion of valid inference. Depending on the semantic notions involved, the reduction of the notions of possibility will be more or less clear and helpful, as you can read in my *Propositional Logics*.

With propositional logics we have analyses of possibility and valid inference for reasoning in the restricted context of paying attention to only propositions as wholes and ways to build propositions from them in a regular way using certain connectives. Each analysis, that is, each logic, reduces the notion of possibility to what are claimed to be simpler notions, which includes the truth or falsity of atomic propositions. But we are far from a sufficient analysis of possibility. Consider:

- All men are mortal. Socrates is a man. Socrates is not mortal.
- All dogs are brown. Some dogs are green.
- Snow is white. Anything that's white isn't black. Snow is black.

We know that each of these collections of propositions is not a description of a way the world could be because we know, informally, that each is or leads to a contradiction. But we cannot show that any of them entails a contradiction if we are restricted to considering the forms of propositions relative to only propositional connectives. We need a way to understand possibilities that takes into account the internal structure of what are atomic propositions in propositional logic.

There are several ways people have chosen to parse the internal structure of propositions in analyses of reasoning. Each depends on a particular view of the world. In this text we based our analysis on the assumption that the world is made up, at least in part, of individual things, and that we would consider reasoning only if it could be construed as being about individual things.

A possibility in predicate logic, then, is a way to assign truth-values to the atomic propositions in a realization. Those include wffs such as "($-$ is a dog) (x_1)" when a reference is supplied for "x_1". So not only truth-values are assumed but also some idea of naming or picking out references from a specified collection of things, what we call the "universe" of a realization. That collection cannot include all things, for as we saw, we do not have a clear enough idea of what we mean by "individual thing" to reason without specifying what kind of things we're talking about.

The two semantic notions truth-value and naming plus the evaluation of the connectives and the quantifiers are all we need to understand possibilities in predicate logic, once we have agreed on the formalization of any particular inference. We have seen many examples where this understanding of possibilities correlates with how we informally analyze inferences that can be formalized in this logic. Here, as for propositional logic, any choice of assignments of truth-values to atomic predications must be allowed in order to have a rich enough notion of possibility in this logic. We must allow a model in which "Marilyn Monroe was a man" and "Napoleon lived in California" are taken as true. This is not to say they are true; rather, a possibility is a model, a description (in terms of the linguistic elements we recognize) of a way the world could be, and that is reduced to an assignment of truth-values to atomic predications.

Does classical predicate logic really give us a reduction of the notion of valid inference? To show that an inference is valid we have to consider all possible models of a particular semi-formal language. Often this can be done by considering general properties of the models. However, we often reason without specifying each and every object in the universe, as when we reason about all the pigs in Denmark. We rely on our notion of naming: given that pig, we could name her "x_1" and then "x_1 is black" is true or false. We can claim that the reduction of possibility and valid inference here is simpler and clearer than the full informal notion in any particular case only to the extent that we believe our notion of naming is simpler and clearer. We have made some progress. In terms of the semantic and syntactic assumptions of this logic, what is logically possible and what is a valid inference have been reduced to just the truth-values of atomic propositions and our notion of naming, along with our evaluation of the connectives and quantifiers.

But it is only progress and hardly the whole story of logical possibilities. We saw examples of inferences that depend on time, or the use of relative adjectives and adverbs, or names that don't refer, which we cannot evaluate in classical predicate logic. Classical predicate logic can be extended to allow for evaluating such reasoning, invoking further semantic notions to which we reduce the notion of possibility, as you can see in my *The Internal Structure of Predicates and Names* and *Time and Space in Formal Logic*.

We also saw that we cannot evaluate reasoning that involves talk of masses or process in classical predicate logic . Such reasoning depends on semantic assumptions that are not compatible with considering propositions to be just about individual things. Formal logics can be devised to evaluate such reasoning, as you can see in my *Reasoning about the World as Process*.

We have made some progress in understanding possibilities and valid inferences and how to reason well. We look at an inference and try to agree on what semantic assumptions we use in informally analyzing it. Then we ask whether we have or can devise a formal logic based on those semantic assumptions. If yes, then we try to formalize the inference in that logic relative to those assumptions. If we are successful, we have clarified in this instance our claim that we do or do not have a logical possibility and whether we do or do not have a valid inference.

In some cases it is quite easy to formalize an inference. In other cases it is quite difficult, requiring us to question our understanding of language and the world. And in other cases there is no formal logic we can use; we have only our informal notions on which to rely. To judge an inference as valid or invalid requires us to be explicit about the semantic assumptions we are making so that we can be clear enough to reason well together. When we can agree, we have made some progress.

We have a beginning. We have tools for how to reason better. We want more: more ways to analyze possibilities in terms of form and meaning, more ways of refining our notion of individual thing, and a way to contrast that conception with other conceptions of the world. There are so many possibilities. I hope to have started you on the journey.

APPENDICES

Appendix 1 Proof by Induction

What's the sum of $1 + 2$? Of $1 + 2 + 3$? That's easy. What's the sum of $1 + 2 + 3 + 4 + 5 + 6 + 7 + 8$? What's the sum of all the counting numbers from 1 to 100? That's hard. But there's a simple way to calculate all those. For any number n,

$$1 + 2 + \cdots + n = 1/2 \, (n \bullet (n + 1))$$

So $\quad 1 + 2 + \cdots + 100 = 1/2 \, (100 \bullet (100 + 1)) = 5{,}050$.

How do we know this formula is right? We can check it for the first few numbers:

$n = 1 \quad 1 = 1/2 \, (1 \bullet (1 + 1))$
$n = 2 \quad 1 + 2 = 1/2 \, (2 \bullet (2 + 1))$
$n = 3 \quad 1 + 2 + 3 = 1/2 \, (3 \bullet (3 + 1))$

But no matter how many numbers we check, that won't show the formula is correct for all numbers n. Rather, we show that if it's correct for some number, then it's correct for the next largest number:

Suppose $\quad 1 + 2 + \cdots + n = 1/2 \, (n \bullet (n + 1))$
then $\quad 1 + 2 + \cdots + n + (n + 1) = [1/2 \, n \bullet (n + 1)] + (n + 1)$
then $\quad 1 + 2 + \cdots + n + (n + 1) = 1/2 \, (n^2 + n) + 1/2 \, (2n + 2)$
then $\quad 1 + 2 + \cdots + n + (n + 1) = 1/2 \, (n^2 + 3n + 2)$
then $\quad 1 + 2 + \cdots + n + (n + 1) = 1/2 \, (n + 1) \bullet (n + 2)$

That is, $1 + 2 + \cdots + n + (n + 1) = 1/2 \, (n + 1) \bullet ((n + 1) + 1)$.

Somehow this looks fishy. After all, we started by supposing that the formula is correct. But no, we started by showing that the formula is correct for $n = 1$. Then we showed that if it is correct for some number n, then it's correct for $n + 1$. So we conclude that it's correct for all numbers. This is the method of proof by *mathematical induction*. Why does it work? Well, the formula is true for 1; so it's true for 2; since it's true for 2, it's true for 3; and so on.

Those little words "and so on" carry a lot of weight. We believe that the counting numbers are completely specified by their method of generation: add 1, starting at 0.

0   1   2   3   4   5   6   7   ...

To prove a statement A by induction, we first prove it for some starting point in this list of numbers, usually 1, but just as well 0 or 47; that's the *basis* of the induction. We then establish that we have a method of generating proofs which is correlated to the method of generating natural numbers: if $A(n)$ is true, then $A(n + 1)$ is true. So we have the list:

A(0)   if A(0), then A(1),   if A(1), then A(2),   if A(2), then A(3),   ...
            so A(1)   so A(2)   so A(3)

Then the statement is true for all natural numbers equal to or larger than our initial point, whether the statement is for all numbers larger than 1 or all numbers larger than 47.

We have really only one idea: a procedure for generating objects one after another without end, in one case numerals or numbers, and in the other, proofs. We believe induction is a correct form of proof because the two applications of the single idea are matched up.

Here's an example in which the basis of the induction is not 0 or 1.

Theorem $1 + 2^n < 3^n$ for $n \geq 2$.

Proof Notice that the statement is false for $n = 0$ and $n = 1$.
  Basis $1 + 2^2 = 5 < 3^2 = 9$.
  Induction step Assume for a fixed $n \geq 2$ that $1 + 2^n < 3^n$.
  Then:
$$\begin{aligned}
1 + 2^{n+1} &= 1 + (2 \cdot 2^n) \\
&= (1 + 2^n) + 2^n \\
&< (1 + 2^n) + (1 + 2^n) \\
&< 3^n + 3^n \quad \text{by the induction hypothesis} \\
&< 3^n + 3^n + 3^n \\
&= 3 \cdot 3^n \\
&= 3^{n+1}
\end{aligned}$$
That is, $1 + 2^{n+1} < 3^{n+1}$, which was to be proved. ∎

We can use induction for any objects we can number. In particular, we can prove theorems about all well-formed-formulas by showing the claim is true for wffs of length 1, and then assuming it is true for wffs of length n, we show it for wffs of length $n + 1$.

Appendix 2 Set-Theory Notation

Some notation is useful to talk about collections and things that are in them.

{a,b,c, ... } stands for the collection consisting of a,b,c, ...

    For example, {Ralph, Howie, Dusty} stands for the collection consisting of Ralph, Howie, and Dusty.

    $\{p_0, p_1, p_2, \dots\}$ stands for the collection of all propositional variables. This is not to assume that there is an infinite collection; it's just a convenient way to talk about all of those.

$\Gamma \cup \Delta$ stands for the collection that contains all the things that are in Γ or that are in Δ, including those that are in both.

    We read $\cup$ as "union".

    For example, {Ralph, Dusty} $\cup$ {Ralph, George} = {Ralph, Dusty, George}.

    $\{p_0, p_1, p_2\} \cup \{p_1, p_3, p_6\} = \{p_0, p_1, p_2, p_3, p_6\}$

$\Gamma \cap \Delta$ stands for the collection that contains all the things that are in both Γ and in Δ.

    We read $\cap$ as "intersection".

    For example, {Ralph, Dusty} $\cap$ {Ralph, George} = {Ralph}.

    $\{p_0, p_1, p_2\} \cap \{p_1, p_3, p_6\} = \{p_1\}$

$a \in \Gamma$ means that a is in Γ.

    We read $\in$ as "is an element of".

    For example, Dusty $\in$ {Ralph, Dusty, George}.

    $p_6 \in \{p_0, p_1, p_2, p_3, p_6\}$

$a \notin \Gamma$ means that a is not in Γ.

    We read $\notin$ as "is not an element of".

    For example, Dusty $\notin$ {Ralph, George}.

$\Gamma \subseteq \Delta$ means that every thing that is in Γ is also in Δ.

    That is, for every a, if $a \in \Gamma$, then $a \in \Delta$.

    We read $\subseteq$ as "is contained in or equal to".

    For example, {Ralph, Dusty} $\subseteq$ {Ralph, Dusty, George}.

    $\{p_0, p_1, p_2, p_3, p_6\} \subseteq \{p_0, p_1, p_2, \dots\}$.

$\{a : \alpha\}$ where α is a condition, stands for the collection of all things satisfying the condition α.

    We read the colon as "such that".

    For example, {a: a is a formal wff that is a conjunction} is a way to talk of all formal wffs that are conjunctions.

    {a: a is a dog and a is brown} is a way to talk of all brown dogs.

Appendix 3 Naming, Pointing, and What There Is

> Agreements 165
> Naming, pointing, and descriptions 166
> Forms of pointing: what there is 168

Agreements

Naming is something we do.

In the simplest case we name by pointing and dubbing. My neighbors get a new dog and say, "Your name is 'Bud'. Understand? 'Bud', your name is 'Bud'." Or a couple has a child and they say, "Let's name him 'Andrew' ", though already here the pointing may be indirect, using words like "him", "the baby", "our child", or a birth certificate to indicate, rather than physically directing our attention to the child.

Pointing and uttering a word (in this simplest case) is how we name or indicate a name: "Juney", "Ralph". But it is also how we say what kind of thing we wish to draw someone's attention to: "table", "chair". It is how we indicate a collection: "herd", "pencils". It is also how we break experience into actions: "running", "licking". And it is how we distinguish the quality of what we encounter: "good", "fast". Only with naming or indicating by use of a name do we not use a generic word. To make that clearer, our language has special words that we understand are used as labels to pick out one and only one thing.

If a foreign student comes to live with me and wants to learn English, I may stick labels on things: "table", "chair", "window". It's generally clear to the student that the labels are generic words, not names, because I use the same word for different objects. It's also clear that the word "Juney" on my dog's collar and the words on name tags worn by students at the language school are names. But it's surprising how often when I teach English a student will look up a name in a dictionary.

An example occurred when a friend and I were visiting his in-laws in Sweden. Neither of us spoke Swedish, but we were eager to learn. An acquaintance offered to show us how to greet someone. He turned to me and said "Tom" while forcefully shaking my hand just once. My friend and I then turned to each other, shook hands, and both said "Tom". Our acquaintance laughed, "No, no. 'Tom' is my name; you're to say yours." The entire action is understood in Sweden as pointing and indicating a name. Customs and background as often as not serve to distinguish the act of naming or picking out a single object with a name from the other ways of pointing and uttering a word.

When I point to something, I hope you'll know what object I'm pointing to. But that is always and only a hope. "That," I say, and you know I mean the table. "That," I say, and you think I mean the pencil on my desk whereas I was pointing to all the things on my desk. And sometimes I don't properly know what I'm pointing to: "That," I say, and think I'm pointing to a statue of a rabbit and its baby, but it moves and is not one thing but two: a rabbit and its baby. Or I point and say "that" in the dark where I heard a noise. Or you point and say "gavagai" and I try to discover whether you mean the rabbit, or a part of the rabbit, or the act of the rabbit running, or . . . , and perhaps those questions never occurred to you. You might not break up experience that way, so by my standard you don't even know what you're pointing to.

So long as we speak the same language we can direct each other's attention to what we are pointing to by using, for example, common nouns. I say, "that table," and I can be pretty sure you'll understand what object I'm pointing to. I point and say, "all those things on my desk," and there's still a good chance you won't misconstrue me. But do you really understand me? All we have is language and more language to explain that, and more language still, and our actions. And our faith that the world is the same for you as for me. And our faith that there *is* a world corresponding in some essential way to our language. That faith may be well-placed, but we'll never know. And though it may be well-placed in its general outlines—there is a world; you perceive objects as I do; . . .—it could be wrongly assumed in many if not all particulars: the word "fork" picks out different objects for you than for me though we agree on the majority of them; your conception of all things on my table might be far different from mine.

There is a spectrum of confidence in our communication. I feel certain that you understand what I am pointing to when I say "that table". I feel less certain when I say "that rutabaga", and quite bewildered when you say "gavagai". The spectrum goes from certainty because of our common background and language to only a hope because of differences in our perceptions and ways of acting as well as our experiences. The certainty may be well-placed for physical or metaphysical reasons; the hope may be less well-founded. But how are we to know with certainty that our belief in certainty is well-founded? We need not resolve this to do logic. We need only agree. The hope, the faith, the belief, the certainty may lie behind our agreements, but we need not examine them.

We establish a semi-formal language and agree that propositions have truth-values and names have reference. We may then further agree that this is the reference of that name, and this proposition is true, that false. The processes are similar, and likely not independent: they both are on the spectrum of confidence that our words connect to "the world". They are agreements *about* the semi-formal language but not *in* the semi-formal language.

Naming, pointing, and descriptions

So when we point and utter a word, we hope to be understood. And that hope lies behind communication. That hardly seems worth mentioning. Except for names.

When I point and say "table" you'll understand what I'm pointing at. You've had enough examples to generalize, you've used the word enough. I do not know how we come to generalize. Perhaps it's because there is an ideal table to which we make comparisons through our intellect; or perhaps it's because we are built so similarly, and the nature of our bodies leads us to see in the same way; or perhaps . . . . There is no end to the explanations, and I am at a loss to understand what it would mean for one of them to be "right". But we do, generally it seems, understand the use of common nouns.

When I point and say "Juney", however, I have used a name, a word that is to pick out only one thing. You recognize the word as a name, but as a name of *what*? There are few, and no reliable, clues in our language: the name suggests a female, probably a living creature, but it may be used otherwise. In the direction I'm pointing are a man, a dog, a woman, a car, a tree. Is one of these Juney?

Appendix 3 Naming, Pointing, and What There Is

When I say "Juney" I mean to pick out one and only one thing. Better would be if I could actually place a label on that object, though even then you could misconstrue what is being labeled. So I direct your attention to the thing by using other language from our common stock of generic words that you are more likely to understand as I do.

I say "Juney is that dog", and "Juney is the one standing near the house", and "Juney is black and white", and "Juney is the small dog". Eventually, I suspect, you'll know what object I'm referring to. The descriptions help me point as much as if I went up to Juney, brought her over to you, lifted her up, and said, "This is Juney".

But there's a big difference between bringing Juney over and lifting her up for you to see, the ultimate in pointing, and using descriptions to point to her. A name is a label, and when I pick up Juney and say, "This is Juney", well, that's Juney. No doubt, except for the always threatening possibility that you think I've named her midriff or the act of her licking you. But that's background, and we're assuming we share that.

Descriptions are a way to avoid the threat of misunderstanding. "The dog," I say, "the dog is Juney". I clarify for you, perhaps completely, what object is named. If not, I add more descriptions based on our commonly understood language until your attention is directed to just that one thing. But unlike labeling, a description may be wrong. Indeed, a description may be wrong even while directing your attention to the right object. "Caesar, who's Caesar?" "Oh, he's the Roman emperor who crossed the Rubicon with his army." You know well enough who I mean, even though Caesar was never emperor.

Or consider that I'd like to get your agreement to use certain words as names in this text:

Ralph	the small purple cloth dog hand puppet given to me by S. and M. Krajewski, as in the picture on the back cover of this book
Juney	the black and white female border collie mix that belonged to my neighbor in Berkeley but who thought she belonged to me and went for a walk with me every day for two and a half years until she was killed by a hit-and-run driver
Arf	me, the author of this book
Marilyn Monroe	the blonde American actress of great fame in the 1950s and 1960s who was married to Arthur Miller and who committed suicide

The description I've attached to a name here does not replace the name. It helps us find the named object. It's a form of pointing.

I could add photos; that's another form of pointing. But we don't want photos or gestures to replace names either. Besides, what photo could I provide of Caesar? No form of pointing should replace a name.

I don't care how the name got reference: what kind of pointing originally accomplished the naming or convinces us now that there's an object denoted by the name. It's enough that the name now has reference. The original pointing that accomplished the naming may have long been forgotten, so that each of us, distant from the origin, associates a different way of pointing to establish reference for that name now. Such ways of pointing constitute (in part) the connotation of a name.

168 An Introduction to Formal Logic

There are many ways to point. All that matters now is that we can agree that a name has reference. If a name has no reference, then we are left with only the pointing, the connotation. By our agreement that names refer, we haven't investigated that possibility here.

Forms of pointing: what there is
Naming, pointing, and what there is are not independent. What we accept as objects is determined in part by (or determines) how we understand naming and reference.

For example, how many things are on my desk? Intuitively we think of things big enough for us to grasp physically and which are unitary. But consider:

Is this one thing? How many parts does it have? Is each a *thing*?

A simpler example:

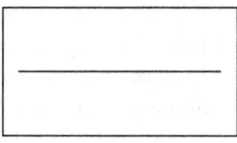
Sam

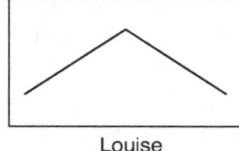
Louise

How many things are in the boxes? The intuitive response is, I believe:

In Sam: 1 (a line)
In Louise: 1 or 2 (2 segments? 1 line?)

Let's construe the diagrams not as representations of abstract lines but as actual physical parts. Now imagine my pointing rather than writing on the diagram, or better, imagine there's a clear plastic sheet with letters and dots on it that I can put over these boxes that makes them look like:

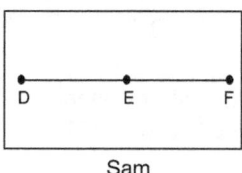
Sam

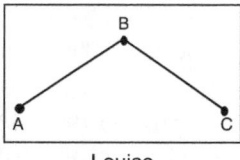
Louise

How many things are in Sam? It's still the same box.

Or consider:

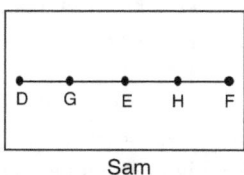
Sam

Now how many things are in Sam? Do we have more line segments? Continue on—are there infinitely many line segments, infinitely many points (real physical points)?

What things we say are in Sam, what we classify as a thing, seems to depend on what we pay attention to or what our attention is drawn to, not simply on what kind of thing we are discussing, though that, too, is important. Or perhaps only what kind of thing is important: how many *labeled* line segments? But in Louise no line segment was originally labeled. How many implicit line segments?

It's not what our attention is drawn to now but what we say our attention can be drawn to: what pointing we accept, what ways we accept for providing references for variables. Generally, saying what kind of thing we are describing is enough. We rely on the "obvious", our intuitions, clarified by specific agreements when necessary, in determining what things there are and how we distinguish them.

It's what we pay attention to and what we choose to ignore that is significant in reasoning. Even if you think there is one absolute, overarching notion of individual thing, we don't (or can't?) use it. Agreements on what things, perhaps what kind of things we are discussing, and on what we shall count as pointing, are essential. Different stories of pointing lead not only to different ideas of what exists but to different ways of how to reason about what exists: different logics. But that's another story.

Appendix 4 Completeness Proofs

    Classical Propositional Logic (PC) 170
    Classical Predicate Logic 174
    Classical Predicate Logic with Equality 180

"Axiom" stands for "axiom scheme".
When no proof is given, I've left to you to provide a proof.

Classical Propositional Logic, PC
The language is L is $L(\neg, \rightarrow, \wedge, \vee, p_0, p_1, \ldots)$.
Every wff that is an instance of one of the following schemes is an axiom.

1. $\neg A \rightarrow (A \rightarrow B)$
2. $B \rightarrow (A \rightarrow B)$
3. $(A \rightarrow B) \rightarrow ((\neg A \rightarrow B) \rightarrow B)$
4. $(A \rightarrow (B \rightarrow C)) \rightarrow ((A \rightarrow B) \rightarrow (A \rightarrow C))$
5. $A \rightarrow (B \rightarrow (A \wedge B))$
6. $(A \wedge B) \rightarrow A$
7. $(A \wedge B) \rightarrow B$
8. $A \rightarrow (A \vee B)$
9. $B \rightarrow (A \vee B)$
10. $(A \rightarrow C) \rightarrow ((B \rightarrow C) \rightarrow ((A \vee B) \rightarrow C))$

rule $\dfrac{A, A \rightarrow B}{B}$ modus ponens

Lemma 1 Soundness of the axiomatization
    If $\Gamma \vdash A$, then $\Gamma \vDash A$.

Lemma 2   a. $\vdash A \rightarrow A$
            b. $\{A, \neg A\} \vdash B$
Proof These are shown in Chapter 6, p. 44 and p. 46.

Lemma 3 The Syntactic Deduction Theorem
    a. $\Gamma, A \vdash B$ iff $\Gamma \vdash A \rightarrow B$
    b. $\Gamma \cup \{A_1, \ldots, A_n\} \vdash B$ iff $\Gamma \vdash A_1 \rightarrow (A_2 \rightarrow (\cdots \rightarrow (A_n \rightarrow B) \cdots))$
Proof (a) If $\Gamma \vdash A \rightarrow B$, then by *modus ponens*, $\Gamma, A \vdash B$.
    To show that if $\Gamma, A \vdash B$, then $\Gamma \vdash A \rightarrow B$, suppose that $B_1, \ldots, B_n$ is a proof of B from $\Gamma \cup \{A\}$. I'll show by induction that for each i with $1 \leq i \leq n$, $\Gamma \vdash A \rightarrow B_i$.
    Either $B_1 \in \Gamma$ or B_1 is an axiom, or B_1 is A. In the first two cases the result follows by using Axiom 2. If B is A, it follows by Lemma 2.a.

Now suppose for all $k < i$, $\vdash A \to B_k$. If B_i is an axiom, or $B_i \in \Gamma$, or B_i is A, we have $\vdash A \to B_i$ as before. The other case is when B_i is a consequence by *modus ponens* of B_m and $B_j = B_m \to B_i$ for some $m, j < i$. By induction we have $\Gamma \vdash A \to (B_m \to B_i)$ and $\Gamma \vdash A \to B_m$, so by Axiom 4, $\Gamma \vdash A \to B_i$. ∎

Lemma 4
    a. Γ is consistent iff there is some B such that $\Gamma \nvdash B$.
    b. $\Gamma \cup \{A\}$ is consistent iff $\Gamma \nvdash \neg A$.
    c. $\Gamma \cup \{\neg A\}$ is consistent iff $\Gamma \nvdash A$.
    d. If Γ is consistent, then $\Gamma \cup \{A\}$ or $\Gamma \cup \{\neg A\}$ is consistent.
    e. If Γ is consistent and complete, then Γ is a theory.
    f. Γ is consistent iff every finite subset of Γ is consistent.

Proof (a) We prove the contrapositives. If Γ is not consistent, then for some A, $\Gamma \vdash A$ and $\Gamma \vdash \neg A$. So by Lemma 2.b, for every B, $\Gamma \vdash B$.

If $\Gamma \vdash B$ for every B, then $\Gamma \vdash p_1$ and $\Gamma \vdash \neg p_1$, so Γ is inconsistent.

(b) We prove the contrapositives. Suppose $\Gamma \vdash \neg A$. So $\Gamma \cup \{A\} \vdash \neg A$. But also $\Gamma \cup \{A\} \vdash A$. So $\Gamma \cup \{A\}$ is inconsistent. Now suppose that $\Gamma \cup \{A\}$ is inconsistent. By part (a), $\Gamma \cup \{A\} \vdash \neg A$. Hence by Lemma 3, $\Gamma \vdash A \to \neg A$. By Lemma 2.a, $\Gamma \vdash \neg A \to \neg A$. Hence by Axiom 3, $\Gamma \vdash \neg A$.

(c) The proof is as for (b).

(d) For the contrapositive, suppose both $\Gamma \cup \{A\}$ and $\Gamma \cup \{\neg A\}$ are inconsistent. Then by (b) and (c), $\Gamma \vdash A$ and $\Gamma \vdash \neg A$, and hence Γ is inconsistent.

(e) Suppose that Γ is complete and consistent, and $\Gamma \vdash A$. If A is not in Γ, then $\neg A$ is in Γ, and hence $\Gamma \vdash \neg A$, so that Γ would be inconsistent. So A is in Γ.

(f) We'll show the contrapositives. If Γ is inconsistent, then for some A, $\Gamma \vdash A$ and $\Gamma \vdash \neg A$. Let $B_1, \ldots, B_n$ be a proof of A from Γ, and let $C_1, \ldots, C_m$ be a proof of $\neg A$ from Γ. Then we have that $\{B_1, \ldots, B_n, C_1, \ldots, C_m\}$ is a finite subset of Γ that is not consistent. In the other direction, if some finite subset $\Delta \subseteq \Gamma$ is inconsistent, then for some A, $\Delta \vdash A$ and $\Delta \vdash \neg A$. But the same proofs are proofs from Γ, too. So Γ is inconsistent. ∎

Lemma 5 Γ is a complete and consistent theory iff there is a model with valuation v such that $\Gamma = \{A: v(A) = \top\}$.

Proof I'll let you show that the collection of wffs true in a model is a complete and consistent theory.

Suppose now that Γ is a complete and consistent theory. Define a valuation by:

$v(p) = \top$ iff $p \in \Gamma$ for every propositional variable p

I'll prove by induction on the length of wffs that $v(A) = \top$ iff $A \in \Gamma$.

It is true for wffs of length 1 by definition. So suppose that's true for all wffs of length $\leq n$ and A and B have length $\leq n$.

$v(\neg A) = \top$ iff $v(A) = \mathsf{F}$
                 iff $A \notin \Gamma$
                 iff $\neg A \in \Gamma$ because Γ is complete and consistent

$v(A \wedge B) = T$ iff $v(A) = T$ and $v(B) = T$
　　　　　　 iff $A \in \Gamma$ and $B \in \Gamma$
　　　　　　 iff $(A \wedge B) \in \Gamma$ using Axioms 5, 6 and 7

$v(A \rightarrow B) = T$ iff $v(A) = F$ or $v(B) = T$
　　　　　　 iff $A \notin \Gamma$ or $B \in \Gamma$
　　　　　　 iff $\neg A \in \Gamma$ or $B \in \Gamma$　because Γ is complete
　　　　　　 iff $(A \rightarrow B) \in \Gamma$ using Axioms 1 and 2

If $A \vee B \in \Gamma$, suppose by way of contradiction that $v(A \vee B) = F$. Then $v(A) = F$ and $v(B) = F$. So by induction, $A \notin \Gamma$ and $B \notin \Gamma$. Since Γ is complete, $\neg A \in \Gamma$ and $\neg B \in \Gamma$. So by Axiom 1, for every C, $\Gamma \vdash A \rightarrow C$ and $\Gamma \vdash B \rightarrow C$. So by Axiom 10, for every C, $\Gamma \vdash A \vee B \rightarrow C$. Since $A \vee B \in \Gamma$, it follows by Lemma 4.a that Γ is inconsistent, a contradiction. So, $v(A) = T$ or $v(B) = T$. So $v(A \vee B) = T$,

If $v(A \vee B) = T$, then $v(A) = T$ or $v(B) = T$. So $A \in \Gamma$ or $B \in \Gamma$. So by Axioms 8 and 9, $A \vee B \in \Gamma$.

Hence, for all A, $v(A) = T$ iff $A \in \Gamma$. ∎

Lemma 6　a. If $\nvdash D$, then there is some complete and consistent theory Γ such that $D \notin \Gamma$.
　　b. Σ has a model iff Σ is consistent.
　　c. If $\Sigma \nvdash A$, then $\Sigma \cup \{\neg A\}$ has a model.

Proof (a) Let $A_0, A_1, \ldots$ be a numbering of the wffs of the formal language (see *Computability* by Walter Carnielli and me). Define:

$\Gamma_0 = \{\neg D\}$

$\Gamma_{n+1} = \begin{cases} \Gamma_n \cup \{A_n\} & \text{if this is consistent} \\ \Gamma_n & \text{otherwise} \end{cases}$

$\Gamma = \bigcup_n \Gamma_n$

We have that Γ_0 is consistent by Lemma 4.b. So by construction, each Γ_n is consistent. Hence Γ is consistent, for if not some finite $\Delta \subseteq \Gamma$ is inconsistent by Lemma 4, and Δ being finite, $\Delta \subseteq \Gamma_n$ for some n, so Γ_n would be inconsistent.

Γ is complete: if $\Gamma \nvdash A$, then by Lemma 4, $\Gamma \cup \{\neg A\}$ is consistent, and hence by construction, $\neg A \in \Gamma$, and so $\Gamma \vdash \neg A$.

Finally, Γ is a theory: if $\Gamma \vdash A$, then since Γ is consistent, $\Gamma \cup \{A\}$ is consistent, and hence by construction, $A \in \Gamma$. ∎

(b) If Σ has a model, it is consistent. If Σ is consistent, then by Lemma 4.a, for some D, $\Sigma \nvdash D$. In the construction of part (a), take $\Gamma_0 = \Sigma \cup \{\neg D\}$, and we get a complete and consistent theory Γ such that $\Sigma \subseteq \Gamma$. S by Lemma 5, Γ has a model.

(c) If $\Sigma \nvdash A$, then by Lemma 4.b, $\Sigma \cup \{\neg A\}$ is consistent. So by part (b), $\Sigma \cup \{\neg A\}$ has a model.

Theorem 7
　　a. *Completeness*　$\Gamma \vdash A$ iff $\Gamma \vDash A$.
　　b. *Compactness*　Γ has a model iff every finite subset of Γ has a model.
　　c. If $\vdash A$, then there is a proof of A in which the only propositional variables that appear in the proof are those that appear in A.

Proof a. By Lemma 1, we need only show that if $\Gamma \models A$ then $\Gamma \vdash A$.

If $\Gamma \models A$, suppose that $\Gamma \nvdash A$. Then by Lemma 6.c, $\Gamma \cup \{\neg A\}$ has a model. So $\Gamma \nvDash A$, a contradiction. Hence, $\Gamma \vdash A$.

b. Γ has a model iff Γ is consistent (Lemmas 5 and 6). Γ is consistent iff every finite subset of Γ is consistent (Lemma 4). Every finite subset of Γ is consistent iff every finite subset of Γ has a model (Lemmas 5 and 6).

c. All the proofs here apply regardless of the collection of propositional variables in the formal language. If $\vdash A$, then $\models A$. Hence, A is valid also for the language whose only propositional variables are those that appear in A. Hence, by part (a), A is a theorem of the language whose only propositional variables are those that appear in A.

In the proof of Lemma 6, the definition of Σ requires that we view the collection of consequences of a collection of wffs as a completed infinity that we can survey at one time or else that we can complete an infinite search to determine whether for every B, not both $\Gamma \vdash B$ and $\Gamma \vdash \neg B$. In *Computability*, Walter Carnielli and I present a proof of the completeness theorem that does not invoke any infinitistic assumptions: given a tautology a proof is constructed.

Classical Predicate Logic

- The language L is $L(\neg, \rightarrow, \wedge, \vee, \forall, \exists; P_0, P_1, \ldots, c_0, c_1, \ldots)$.
- "PC" is an abbreviation for "classical propositional logic".

In the proofs below I write $\sigma \vDash A$ as an abbreviation of $\nu_\sigma \vDash A$.

I. *Propositional axioms*
   (the axiom schemes of classical propositional logic)

   $\forall \ldots [\neg A \rightarrow (A \rightarrow B)]$

   $\forall \ldots [B \rightarrow (A \rightarrow B)]$

   $\forall \ldots [(A \rightarrow B) \rightarrow ((\neg A \rightarrow B) \rightarrow B)]$

   $\forall \ldots [(A \rightarrow (B \rightarrow C)) \rightarrow ((A \rightarrow B) \rightarrow (A \rightarrow C))]$

   $\forall \ldots [A \rightarrow (B \rightarrow (A \wedge B))]$

   $\forall \ldots [(A \wedge B) \rightarrow A]$

   $\forall \ldots [(A \wedge B) \rightarrow B]$

   $\forall \ldots [A \rightarrow (A \vee B)]$

   $\forall \ldots [B \rightarrow (A \vee B)]$

   $\forall \ldots [(A \rightarrow C) \rightarrow ((B \rightarrow C) \rightarrow ((A \vee B) \rightarrow C))]$

II. *Axioms governing* $\forall$

   1. a. $\forall \ldots (\forall x (A \rightarrow B) \rightarrow (\forall x A \rightarrow \forall x B))$ *distribution of* $\forall$
         if x is free in both A and B
      b. $\forall \ldots (\forall x (A \rightarrow B) \rightarrow (\forall x A \rightarrow B))$
         if x is free in A and not free in B
      c. $\forall \ldots (\forall x (A \rightarrow B) \rightarrow (A \rightarrow \forall x B))$
         if x is free in B and not free in A

   2. $\forall \ldots (\forall x \forall y A \rightarrow \forall y \forall x A)$ *commutativity of* $\forall$

   3. $\forall \ldots (\forall x A(x) \rightarrow A(c/x))$ *universal instantiation*

III. *Axioms governing the relation between* $\forall$ *and* $\exists$

   4. a. $\forall \ldots (\exists x A \rightarrow \neg \forall x \neg A)$
      b. $\forall \ldots (\neg \forall x \neg A \rightarrow \exists x A)$

   rule $\dfrac{A, A \rightarrow B}{B}$ for A, B closed wffs

Note that only closed wffs are theorems.

Lemma 8 Soundness of the axiomatization   If $\Gamma \vdash A$, then $\Gamma \vDash A$.
This is proved in Chapter 13 (p. 103).

Lemma 9 The Syntactic Deduction Theorem
   a. $\Gamma, A \vdash B$ iff $\Gamma \vdash A \to B$.
   b. $\Gamma \cup \{A_1, \ldots, A_n\} \vdash B$ iff $\Gamma \vdash A_1 \to (A_2 \to (\cdots \to (A_n \to B) \cdots))$.

Proof a. Since A, B, and all wffs in Γ are closed, the proof is as for PC, p. 170.
   b. By induction on (a). ∎

Lemma 10 If $\vdash A \to B$ and $\vdash B \to C$, then $\vdash A \to C$.

Proof If $\vdash A \to B$ and $\vdash B \to C$, then $A \vdash B$ and $B \vdash C$. Combining those proofs we have a proof of C from A. So by the Syntactic Deduction Theorem, $\vdash A \to C$. ∎

Lemma 11 a. $\vdash \forall \ldots (A \to B) \to (\forall \ldots A \to \forall \ldots B)$
   b. Generalized *modus ponens* $\{\forall \ldots (A \to B), \forall \ldots A\} \vdash \forall \ldots B$

Proof a. We proceed by induction on the number of variables in the universal closure for $(A \to B)$. If $n = 0$, this is just $\vdash (A \to B) \to (A \to B)$, which we have by the PC axioms as in Lemma 2. If $n = 1$, this is one of the schemes of Axiom 1. Suppose the lemma is true for n. Let x be the last variable (in alphabetical order) that appears free in $(A \to B)$. We have three cases, depending on whether x appears free in A, B, or both. I will do the case where x is free in A and not in B and leave the others to you. An instance of Axiom 1 is:

$$\forall \ldots (\forall x (A \to B) \to (\forall x A \to B))$$

Since the universal closure of this formula has only n variables, we have by induction $(\forall \ldots (\forall x A \to B) \to (\forall \ldots \forall x A \to \forall \ldots B)$, which was to be proved. ∎

Lemma 12 PC *in predicate logic* If A has the form of a PC-tautology, then $\vdash \forall \ldots A$.

Proof Suppose that one form of A is the PC-tautology B. By the completeness theorem for PC, there is a proof $B_1, \ldots, B_n = B$ in PC. By Theorem 7.c, we can assume there is no propositional variable appearing in any of $B_1, \ldots, B_n$ that does not also appear in B. Let B_i^* be B_i with propositional variables replaced by predicate wffs just as they are replaced in B to obtain A. I'll show by induction on i that $\vdash \forall \ldots B_i^*$. The lemma will then follow for $i = n$.

For $i = 1$, B_1 is a PC-axiom. Suppose now that for all $j < i$, $\vdash \forall \ldots B_j^*$. If B_i is a PC-axiom we are done. If not, then for some $j, k < i$, B_j is $B_k \to B_i$. Since $(B_k \to B_i)^*$ is $B_k^* \to B_i^*$, by induction we have $\vdash \forall \ldots B_k^*$ and $\vdash \forall \ldots (B_k^* \to B_i^*)$. Hence by generalized *modus ponens*, $\vdash \forall \ldots B_i^*$. ∎

When invoking Lemma 11 or Lemma 12, I'll just say "by PC". In particular, we have the transitivity of $\to$, $\forall \ldots (A \to B), \forall \ldots (B \to C) \vdash \forall \ldots (A \to C)$.

Lemma 13 Permuting universal quantifiers
   If $y_1 \ldots y_n$ are the variables free in A in alphabetic order, then:
   $\vdash \forall \ldots A \to \forall y_k \forall y_1 \ldots \forall y_{k-1} \forall y_{k+1} \ldots \forall y_n A$.

Proof We induct on n. For $n = 0$ or 1, this is by PC. For $n = 2$ this is an instance of Axiom 2. Suppose now the lemma is true for fewer than n variables. If $k \neq n$, we are done by induction (replacing A by $\forall y_n A$). If $k = n$, then by Axiom 2:

$$\vdash \forall \ldots ((\forall y_{n-1} \forall y_n A) \to (\forall y_n \forall y_{n-1} A))$$

176   An Introduction to Formal Logic

Since in the prefix $\forall \ldots$ the variables appear in alphabetic order, we have:
$$\vdash \forall y_1 \ldots \forall y_{n-2} (\forall y_{n-1} \forall y_n A) \rightarrow \forall y_1 \ldots \forall y_{n-2} (\forall y_n \forall y_{n-1} A)$$
By induction we have:
$$\vdash \forall y_1 \ldots \forall y_{n-2} \forall y_n (\forall y_{n-1} A) \rightarrow \forall y_n \forall y_1 \ldots \forall y_{n-2} (\forall y_{n-1} A)$$
Hence, by Lemma 10, we are done. ∎

Lemma 14   Let $B(x)$ be a formula with one free variable x.
  a. If $\vdash B(c/x)$, then $\vdash \forall x\, B(x)$.
  b. If $\Gamma \vdash B(c/x)$ and c does not appear in any wff in Γ, then $\Gamma \vdash \forall x\, B(x)$.

Proof a. We proceed by induction on the length of a proof of $B(c/x)$. If the proof has length 1, then $B(c/x)$ is an axiom. I will show that (a) holds for instances of Axiom 1.a. Instances of the other Axioms follow similarly.
  One instance of the scheme is:
$$\vdash (\forall \ldots)_1 (\forall y\, (C(c/x) \rightarrow D(c/x))) \rightarrow (\forall y\, C(c/x) \rightarrow \forall y\, D(c/x))$$
But another instance of Axiom 1 is:
$$\vdash (\forall \ldots)_2 (\forall y\, (C(x) \rightarrow D(x))) \rightarrow (\forall y\, C(x) \rightarrow \forall y\, D(x))$$
where the only difference between $(\forall \ldots)_1$ and $(\forall \ldots)_2$ is that $\forall x$ appears in the latter. Hence, by Lemma 13, we are done.
  Suppose now that (a) is true for theorems with proofs of length m, $1 \leq m < n$, and the shortest proof of $B(c/x)$ has length n. Then for some closed A, $\vdash A$ and $\vdash A \rightarrow B(c/x)$, both of which have proofs shorter than length n. By induction, $\vdash \forall x\, (A \rightarrow B(x))$, so by Axiom 1.c and PC, $\vdash A \rightarrow \forall x\, B(x)$. Hence, $\vdash \forall x\, B(x)$.
  b. Suppose $\Gamma \vdash B(c/x)$. Then for some closed $D_1, \ldots, D_n \in \Gamma$, we have that $\{D_1, \ldots, D_n\} \vdash B(c/x)$. Hence by the syntactic deduction theorem:
$$\vdash D_1 \rightarrow (D_2 \rightarrow \cdots \rightarrow (D_n \rightarrow B(c/x)) \ldots)$$
Since by hypothesis c does not appear in any of $D_1, \ldots, D_n$, we have by (a):
$$\vdash \forall x\, (D_1 \rightarrow (D_2 \rightarrow \cdots \rightarrow (D_n \rightarrow B(x)) \ldots))$$
So by Axiom 1, $\vdash D_1 \rightarrow (D_2 \rightarrow \cdots \rightarrow (D_n \rightarrow \forall x\, B(x)) \ldots)$. By repeated use of *modus ponens*, $\{D_1, \ldots, D_n\} \vdash \forall x\, B(x)$. So $\Gamma \vdash \forall x\, B(x)$. ∎

Theorem 15   Let Γ be a consistent set of closed wffs of L. Let $L(w_0, w_1, \ldots)$ be L with the addition of name symbols $w_0, w_1, \ldots$ that do not appear in L. Then there is a collection of closed wffs Σ in $L(w_0, w_1, \ldots)$ such that:
  a. $\Gamma \subseteq \Sigma$.
  b. Σ is a complete and consistent theory.
  c. If $\exists x B \in \Sigma$ and x is free in B, then for some m, $B(w_m/x) \in \Sigma$.
  d. For every wff $B(x)$ in $L(w_0, w_1, \ldots)$ with one free variable,
     if for each i, $B(w_i/x) \in \Sigma$, then $\forall x\, B(x) \in \Sigma$.

Proof Let $A_0, A_1, \ldots$ be a numbering of the closed wffs of the expanded language $L(w_0, w_1, \ldots)$. Let $\vdash$ stand for derivations in this language. Define Σ by stages:

$\Sigma_0 = \Gamma$

Σ_{n+1} is defined by cases:

i. If $\Sigma_n \vdash \neg A_n$, then $\Sigma_{n+1} = \Sigma_n \cup \{\neg A_n\}$.

If $\Sigma_n \nvdash \neg A_n$, then:

ii. If A_n is not $\exists x B$, then $\Sigma_{n+1} = \Sigma_n \cup \{A_n\}$.

iii. If A_n is $\exists x\, B$ and w_m is the least w_i that does not appear in Σ_n, then $\Sigma_{n+1} = \Sigma_n \cup \{\exists x B, B(w_m/x)\}$.

$\Sigma = \bigcup_n \Sigma_n$

Part (a) follows by construction.

For (b), I'll first show by induction that for each n, Σ_n is consistent. We have $n = 0$ by hypothesis. Suppose now that Σ_n is consistent. If Σ_{n+1} is defined by (i), it's immediate that Σ_{n+1} is consistent. If Σ_n is defined by (ii), Σ_{n+1} is consistent follows by induction and Lemma 4. So suppose Σ_{n+1} is defined by (iii). Then $\Delta = \Sigma_n \cup \{\exists x\, B(x)\}$ is consistent by Lemma 4. Suppose now that Σ_{n+1} is not consistent. Then by Lemma 4, $\Delta \vdash \neg B(w_m/x)$. So by Lemma 13, $\Delta \vdash \forall x \neg B(x)$. Hence, by Axiom 4.a and PC, $\Delta \vdash \neg \exists x\, B(x)$. But then Δ is not consistent, which is a contradiction. So Σ_{n+1} is consistent. It then follows that Σ is consistent, for if it were not, then some finite subset of it would be inconsistent, and hence some Σ_n would be inconsistent.

For every A, by construction either $A \in \Sigma$ or $\neg A \in \Sigma$. So Σ is complete, and hence by Lemma 4, Σ is a theory, so (b) is proved.

For (c), suppose $\exists x\, B(x) \in \Sigma$. Then for some n and m,

$$\Sigma_{n+1} = \Sigma_n \cup \{\exists x\, B, B(w_m/x)\} \subseteq \Sigma$$

For (d), I'll show the contrapositive. Suppose $\forall x\, B(x) \notin \Sigma$. Then by (b), $\neg \forall x\, B(x) \in \Sigma$. Hence by Axiom 4.b and PC, $\exists x \neg B(x) \in \Sigma$. So by (c), for some m, $\neg B(w_m/x) \in \Sigma$. So by the consistency of Σ, $B(w_m/x) \notin \Sigma$. ∎

Lemma 16 Existential Generalization

When x is free in A, $\vdash \forall \ldots (A(c/x) \to \exists x A(x))$

Proof By Axiom 3, $\vdash \forall \ldots (\forall x \neg A(x) \to \neg A(c/x))$. So by PC, $\vdash \forall \ldots (A(c/x) \to \neg \forall x \neg A(x))$. So by Axiom 4.b and PC, $\vdash \forall \ldots (A(c/x) \to \exists x A(x))$.

Theorem 17 Every consistent collection of closed wffs in L has a countable model.

Proof Let Γ be a consistent collection of closed wffs of L.

Let $\Sigma \supseteq \Gamma$ in $L(w_0, w_1, \ldots)$ be as in Theorem 15. We define a model M^* of $L(w_0, w_1, \ldots)$. The realization of the formal language is the formal language itself. Set:

$U = \{w_i, c_i : i \geq 0\}$

We take a complete collection of assignments of references such that for every σ,

$\sigma(c_i) = c_i \qquad \sigma(w_i) = w_i$

We evaluate the atomic wffs by:

$v_\sigma \vDash P_i^n(v_1, \ldots, v_n)$ iff $P_i^n(\sigma(v_1), \ldots, \sigma(v_n)) \in \Sigma$

Lemma α Suppose x is free in B and $d \in U$. Then for any σ:
If $\sigma(x) = d$, then $\nu_\sigma \vDash B(x)$ iff $\nu_\sigma \vDash B(d/x)$.

Proof We proceed by induction on the length of wffs. If B has length 1, the lemma follows by definition.

So suppose the lemma is true for all wffs of length $\leq n$ and B has length $n + 1$. I'll leave the cases when B has the form $\neg C, C \rightarrow D, C \wedge D$, or $C \vee B$ to you. So suppose that B has the form $\forall y\, C(x)$. Note that y is not x, for x is free in B. We have:

$\nu_\sigma \vDash \forall y\, C(x, y)$ iff for every τ such that $\tau \sim_y \sigma$, $\nu_\tau \vDash C(x, y)$
 iff for every τ such that $\tau \sim_y \sigma$, $\nu_\tau \vDash C(d/x, y)$
  (by induction, since τ agrees with σ on y)
 iff $\nu_\sigma \vDash \forall y\, C(d/x, y)$

I'll leave to you the case when B has the form $\exists y\, C(y)$.

That the valuations satisfy the extensionality condition follows from this lemma and the definitions of the valuations, since for every $i, j, c_i \neq w_j$.

Now we'll show by induction on the length of wffs that for every closed wff A in $L(w_0, w_1, \ldots)$, $M \vDash A$ iff $A \in \Sigma$.

Suppose A is atomic. Then all terms in A are names, so we are done by the definition of the evaluation of atomic wffs. Now suppose it is true for wffs shorter than A. I'll do the cases when A is $\exists x\, B$ or $\forall x\, B$ and leave the others to you, as in the proof for PC.

Suppose A is $\exists x B$ and $\exists x\, B(x) \in \Sigma$. Then x is the only variable free in $B(x)$. Since Σ satisfies the conditions in Theorem 15, for some m, $B(w_m/x) \in \Sigma$. Hence by induction, for any σ, $\nu_\sigma \vDash B(w_m/x)$. So by Lemma α, for any σ, if $\sigma(x) = w_m$, then $\nu_\sigma \vDash B(x)$. So $M \vDash \exists x\, B(x)$.

Suppose A is $\exists x\, B$ and $M \vDash \exists x\, B(x)$. Then for some σ, $\nu_\sigma \vDash B(x)$. Take one where $\sigma(x) = d$. Then by Lemma α, $\nu_\sigma \vDash B(d/x)$. Since $B(d/x)$ is closed, $M \vDash B(d/x)$, so by induction $B(d/x) \in \Sigma$. By Lemma 16, $\vdash B(d/x) \rightarrow \exists x\, B(x)$. So $\exists x\, B(x) \in \Sigma$.

Suppose A is $\forall x\, B$ and $\forall x\, B(x) \in \Sigma$. Since Σ is a theory, by Axiom 3, for every name d in the language, and hence for all $d \in U$, $B(d/x) \in \Sigma$. Hence by induction, for all $d \in U$, $M \vDash B(d/x)$. Hence by Lemma α, for every σ, $\nu_\sigma \vDash B(x)$, and hence $M \vDash \forall x\, B(x)$.

Suppose A is $\forall x\, B$ and $M \vDash \forall x\, B(x)$. So for every $i \geq 0$, if $\sigma(x) = w_i$, then $\nu_\sigma \vDash B(x)$. So by Lemma α, for every i, for every σ, $\nu_\sigma \vDash B(w_i/x)$. So by induction, for every i, $B(w_i/x) \in \Sigma$. Since Σ satisfies condition (d) of Theorem 15, $\forall x\, B(x) \in \Sigma$.

To complete the proof of the theorem, define a model N for L by taking M* and deleting the interpretation of each w_i (the universes are the same). Then for any closed wff A in L, $M \vDash A$ iff $N \vDash A$, as you can show. So $N \vDash \Gamma$. And N is countable. ∎

Theorem 18
   a. For any collection of closed wffs Σ in L, Σ is a complete and consistent theory iff there is a model M such that $M \vDash A$ iff $A \in \Sigma$.
   b. For any model M of L, there is a countable model M* such that $M^* \vDash A$ iff $M \vDash A$.
   c. *Completeness* $\Gamma \vdash A$ iff $\Gamma \vDash A$.
   d. *Compactness* Γ has a model iff every finite subset of Γ has a model.

Proof a. If $\Sigma = \{A: M \vDash A\}$, then Σ is a complete and consistent theory. The other direction follows by Theorem 17.

b. If M is a model, then $\{A: M \vDash A\}$ is complete and consistent. So by Theorem 17, there is a countable model M* such that $\{A: M \vDash A\} = \{A: M^* \vDash A\}$.

Parts (c) and (d) follow as for Theorem 7. ∎

In the proof of Theorem 15, the definition of Σ requires that we view the collection of consequences of a collection of wffs as a completed infinity that we can survey at one time. Mathematicians have shown that such an infinitistic assumption is essential. For those who do not ascribe to such infinitistic assumptions, it is not proven that every tautology is a theorem.

180   An Introduction to Formal Logic

Classical Predicate Logic with Equality
The following axioms are added to those for classical predicate logic.

Axioms for equality

5. $\forall x \, (x \equiv x)$ *identity*
6. $\forall \ldots \forall x \, \forall y \, (x \equiv y \rightarrow (A(x) \rightarrow A(y/x)))$ *substitution of equals*
   where y replaces some but not necessarily all occurrences of x in A

To prove the strong completeness of this axiomatization, the lemmas and theorems through Theorem 18, along with their proofs, are the same as for classical predicate logic without equality.

Lemma 19   a. $\vdash \forall x \, \forall y \, (x \equiv y \rightarrow y \equiv x)$ *symmetry*
            b. $\vdash \forall x \, \forall y \, \forall z \, (x \equiv y \rightarrow (y \equiv z \rightarrow x \equiv z))$ *transitivity*

Proof a. 1. $\vdash \forall x \, (x \equiv x)$ identity axiom
           2. $\vdash \forall x \, \forall y \, (x \equiv y \rightarrow (x \equiv x \rightarrow y \equiv x))$ substitution of equals
           3. $\vdash \forall x \, \forall y \, (x \equiv x \rightarrow (x \equiv y \rightarrow y \equiv x))$ (2) and PC
           4. $\vdash \forall x \, \forall y \, (x \equiv y \rightarrow y \equiv x)$ (1), (3), and Axiom 1
      b. 1. $\vdash \forall z \, \forall y \, \forall x \, (y \equiv x \rightarrow (y \equiv z \rightarrow x \equiv z))$ substitution of equals
           2. $\vdash \forall x \, \forall y \, (x \equiv y \rightarrow y \equiv x)$ part (a)
           3. $\vdash \forall x \, \forall y \, \forall z \, (x \equiv y \rightarrow (y \equiv z \rightarrow x \equiv z))$ (1), (2), PC, Lemma 11 ∎

Theorem 19   Every consistent set of closed wffs of L has a countable model in which "$\equiv$" is interpreted as the identity of the universe.

Proof Let Γ be a consistent collection of closed wffs of L. Let $\Sigma \supseteq \Gamma$ be constructed as in the proof of Theorem 17. By Theorem 18, there is a countable model M of Σ, and hence of Γ. In that model, "$\equiv$" is evaluated

$$v_\sigma \vDash v \equiv u \text{ iff } (\sigma(v) \equiv \sigma(u)) \in \Sigma$$

We'll write "$\approx$" for this interpretation of "$\equiv$". This need not be the identity on the universe of M. However, as you can show, $\approx$ is an equivalence relation (p. 143) by virtue of Axiom 5, Lemma 19, and Axiom 3. That is, for every c, d, and e in U:

     $c \approx c$
     If $c \approx d$, then $d \approx c$.
     If $c \approx d$, and $d \approx e$, then $c \approx e$.

Denote by [d] the equivalence class of d for $\approx$:

     $[d] = \{ c : c \approx d \}$

We have that $[d] = \{ c : (c \equiv d) \in \Sigma \}$. Given an equivalence class c, we say that d is a *representative* of c if $c = [d]$.

We'll now define a model $M/\approx$ in which "$\equiv$" is interpreted as the identity on the universe such that $M/\approx$ validates exactly the same wffs as M.

The semi-formal language of $M/\approx$ is the same as for M. The universe of $M/\approx$ is:

$\{[d] : d \text{ is in the universe of } \mathsf{M}\}$

For each assignment of references σ of M, define an assignment of references $\sigma/_\simeq(v)$ by setting for every term v,

$$\sigma/_\simeq(v) = [\sigma(v)]$$

Since the collection for M is complete, the collection of these assignments is complete.

For each assignment of references, define the valuation on atomic wffs:

$\mathsf{v}_{\sigma/_\simeq} \vDash P(v_1, \ldots, v_n)$ iff $\mathsf{v}_\sigma \vDash P(v_1, \ldots, v_n)$

$\mathsf{v}_{\sigma/_\simeq} \vDash u \equiv v$ iff $\mathsf{v}_\sigma \vDash u \equiv v$

Hence, by the definition of M,

$\mathsf{v}_{\sigma/_\simeq} \vDash P(v_1, \ldots, v_n)$ iff $P(\sigma(v_1), \ldots, \sigma(v_n)) \in \Sigma$

$\mathsf{v}_{\sigma/_\simeq} \vDash u \equiv v$ iff $(\sigma(v) \equiv \sigma(u)) \in \Sigma$

Now we show that for every closed formal wff A and every σ for M,

$\mathsf{v}_{\sigma/_\simeq} \vDash A$ iff $\mathsf{v}_\sigma \vDash A$

If A has length 1, it is immediate. Suppose now that it is true for all closed formal wffs of length n, and A has length $n + 1$. I will leave the cases when A has the form $\neg$ C, $C \to D$, $C \wedge D$, or $C \vee B$ to you. First note that the choice of representative of the equivalence class does not matter:

($\ddagger$)   For any assignments of references τ, σ, if for all z, $\tau(z) \simeq \sigma(z)$, then by Axiom 6, for every wff A, $\mathsf{v}_\tau \vDash A$ iff $\mathsf{v}_\sigma \vDash A$.

Now suppose that A has the form $\forall x\, B(x)$. We have:

$\mathsf{v}_{\sigma/_\simeq} \vDash \forall x\, B(x)$ iff   for every assignment $\tau/_\simeq \sim_x \sigma/_\simeq$, $\mathsf{v}_{\tau/_\simeq} \vDash B(x)$

iff   for every assignment $\tau \sim_x \sigma$, if $\tau(x) = d$,
then $\mathsf{v}_\tau \vDash B(d/x)$ (by induction and Lemma α and ($\ddagger$))

iff $\mathsf{v}_\sigma \vDash B(x)$ (by Lemma α)

Now suppose that A has the form $\exists x\, B(x)$. We have:

$\mathsf{v}_{\sigma/_\simeq} \vDash \exists x\, B(x)$ iff   for some assignment $\tau/_\simeq \sim_x \sigma/_\simeq$, $\mathsf{v}_{\tau/_\simeq} \vDash B(x)$

iff   for some assignment $\tau \sim_x \sigma$ where $\tau(x) = d$, $\mathsf{v}_\tau \vDash B(d/x)$
(by induction and Lemma α and ($\ddagger$))

iff $\mathsf{v}_\sigma \vDash B(x)$ (by Lemma α)

Hence, for every A, $\mathsf{M} \vDash A$ iff $\mathsf{M}/_\simeq \vDash A$. And in M, "$\equiv$" is evaluated as the identity:

$\mathsf{v}_{\sigma/_\simeq} \vDash v \equiv u$ iff $(\sigma(v) \equiv \sigma(u)) \in \Sigma$
                       iff $\sigma(v) \simeq \sigma(u)$
                       iff $[\sigma(v)] = [\sigma(u)]$

To complete the proof of the theorem, define a model N for L by taking $\mathsf{M}/_\simeq$ and deleting the interpretation of each w_i (the universes are the same). Then for any closed wff A in L, $\mathsf{M}/_\simeq \vDash A$ iff $\mathsf{N} \vDash A$, as you can show. So $\mathsf{N} \vDash \Gamma$. And N is countable. ∎

The completeness and compactness of this axiom system is proved as in Theorem 18.

Appendix 5   Other Interpretations of the Quantifiers and Variables

> The substitutional interpretation 182
> Naming all elements of the universe at once 183
> Surveying all interpretations of the name symbols 184

The substitutional interpretation

The *substitutional interpretation* of the quantifiers restricts the role of variables to just placeholders for names. For example, using a notation that doesn't separate predicates and variables as ours does:

$\forall x_1$ (x_1 is a dog) is true iff no matter what name replaces x_1 in "x_1 is a dog", the resulting proposition is true

A definition of truth in a model is given along the previous lines by deleting any mention of assignments of references and replacing the truth-conditions for quantified formulas by:

(1) $\forall x$ A is true iff no matter what name replaces every free occurrence of x in A, the resulting proposition is true

$\exists x$ A is true iff there is some name such that when every free occurrence of x in A is replaced by that name, the resulting proposition is true

The substitutional interpretation classifies propositions as true or false differently than our definition. For example, according to our definition, "$\exists x_1$ (x_1 is a wombat)" is true for the model described in Example 1 of Chapter 11 (p. 79) with the "obvious" truth-assignments for atomic predications, since we could point to a wombat (say at a zoo) and say, "Let x_1 stand for that". But the proposition is false according to the substitutional interpretation because none of Ralph, Dusty, Howie, or Juney is a wombat. On the substitutional interpretation we can consider only the names we originally took in the realization, for if (1) is meant to allow for other names, we might as well stick with our original analysis and use the variables as temporary names, for then we'd at least agree on what potential names we could use.

But the example is misleading. The substitutional interpretation imposes the obligation to name everything we're talking about *before* we begin reasoning: otherwise $\forall$ and $\exists$ would be poor formalizations of "for all" and "there exists". For a model in which everything in the universe is named, our previous definition and the substitutional definition of truth in a model classify the same sentences as true. The substitutional interpretation should be seen as a restriction on what counts as a realization, not what counts as true.

Is the restriction on realizations appropriate? One cheap answer is to say that just as a logical symbol can be a name for itself, we can consider any object to be a name for itself. Thus the substitutional interpretation doesn't really restrict a realization to objects *we* name. Thanks, but no. I don't want to have to haul some squealing pig up to a piece of paper on which "is a dog" is written in order to show that "$\forall x_1$ (x_1 is a dog)" is false. Use-mention confusions are bad, but they're especially unpleasant if you have to haul a pig around. And short of letting objects name themselves, we have no way to reason about, say, all ducks in New Mexico according to the substitutional interpretation of the quantifiers.

So in general, the restriction to reasoning about only named objects is too much. But in those cases where it is possible to name all objects, the substitutional interpretation is an intuitive alternative to how we previously interpreted quantifiers.

But the substitutional interpretation is a more serious alternative. Consider:

- We don't need to specify a universe for a realization.
  Everything we're talking about is named, and we have the names available to us.
- We don't even need to suppose that the names refer to anything.
  All we need is that truth-values can be assigned to atomic propositions.
- If no universe is postulated, the question of whether two names refer to the same object is moot.
  We have only atomic propositions and truth-values for those, and the question of consistency of predications does not arise, nor does the issue of extensional predications arise.

To interpret the quantifiers substitutionally requires only one semantic primitive: truth-value of atomic propositions. Reference can be dispensed with . In comparison, the definition of satisfaction we gave is called the *referential interpretation* of quantifiers and variables.

Eliminating a semantic primitive seems an advance in defining truth. But we do want to reason about unnamed things. A notion of reference seems if not required by at least an important supplement to the assumption Things, the World, and Propositions. We needn't subscribe to that assumption. We might want to do away with a notion of reference. But then why should we parse propositions as we do in predicate logic? Why should we focus primarily on names in propositions?

Further, on the substitutional interpretation we use the phrases "for every name a" and "for some name a" at the metalevel. Are we to conceive of those quantifications as referential or substitutional? If there are only finitely many names in the realization, the substitutional interpretation will do. But if there are infinitely many names, it seems we have to talk about possible inscriptions, and a substitutional understanding of quantification in the metalanguage seems suspect.

Naming all elements of the universe at once
Joseph Shoenfield in his *Mathematical Logic* (1967) proposes that we act as if it were possible to name each object of the universe. Or as he puts it (p. 18), "For each individual a [of the universe], we choose a new constant, called the name of a." He then expands the semi-formal language to include each of those names and defines "true in the expanded language of the model" via the substitutional interpretation. The predicate "true in the model" is then taken to be the restriction to the original semi-formal language of the predicate "true in the expanded language of the model". He claims that this definition gives the same classification of propositions into true and false as ours.

But to assume that everything in the universe can be named removes names and naming from any basis in our common understanding. Where are we to find names for all real numbers? Shall we have an uncountable language? Even if you're a platonist, names and naming are part of our linguistic usage. And to consider the truth-conditions of "$\forall x_1 \, (x_1$ is a dog)" in the model based on the realization described in Example 1 of Chapter 11.A (p. 79) with the "obvious" truth-assignments for atomic predications, is it really necessary to name all the ducks in New Zealand? Our metalogical analyses would take a very long time.

Though Shoenfield's approach gives a mathematically neat way to do metalogic, replacing talk about names in place of talk about ways of naming is inappropriate to yield a justification for why we should adopt classical predicate logic.

Surveying all interpretations of the name symbols
Benson Mates, in his *Elementary Logic* (1965), offers a different analysis. His goal is to keep the use of names and variables entirely distinct. He eschews the use of variables as pronouns or temporary names: they are placeholders only.

If variables are to be taken as only placeholders, then it would seem that the substitutional interpretation of the quantifiers is what we want. But Mates believes that to be too restrictive.

Instead, he begins with the idea that we should be concerned only with formal languages and their interpretations, avoiding realizations entirely. Then he can consider how a name symbol can be interpreted as an object of the universe while ignoring any realization of it as a name. Thus "c_0" could be interpreted as Ralph, or as Dusty, or as my neighbor's dog Cheema, but not as "Ralph", or "Dusty", or "Cheema". The truth-conditions he gives are:

If x is free in A, let c_j be the first name symbol not appearing in A.

$\forall x \, A$ is true in the model   iff   for any model that differs from the one under consideration only in what it assigns c_j, $A(c_j/x)$ is true

$\exists x \, A$ is true in the model   iff   for some model that differs from the one under consideration only in what it assigns c_j, $A(c_j/x)$ is true

In this way Mates keeps the roles of name symbols and variables distinct. But has he kept the role of names and variables distinct? There is nothing in his formal system that can be construed as a name. What could formalize "Ralph is a dog"? Perhaps $P_0^1(c_0)$? But in evaluating truth in a model we must consider all possible assignments of objects of the universe to c_0 as reference. Where we look at such assignments of references for variables within a model, Mates looks to those assignments via surveying all models that differ from the one under consideration only in what they assign the name symbols. But it amounts to the same: something in the formal language must allow for varying reference. What Mates has done is split the roles we assigned to variables between name symbols, which act as variables in the sense of allowing for varying reference, and variables, which act as placeholders, while deleting entirely the role of name symbols as standing for names.

It is appealing to try to keep the roles of names and variables entirely distinct, using variables as placeholders only. But these attempts to do so either unnecessarily restrict logic, remove names and naming from our common notions, or lead us right back to variables or some other part of the language being used as temporary names.

Appendix 6 Mathematical Semantics

> The Abstraction of Classical Propositional Logic Models . . 185
> The Extension of a Predicate 186
> Mathematical Models of Classical Predicate Logic 189

The Abstraction of Classical Propositional Logic Models

A model of classical propositional logic is:

Type **I**
$$L(\neg, \rightarrow, \wedge, \vee, p_0, p_1, \ldots)$$
$$\downarrow \text{ realization}$$
$$\{\text{real}(p_0), \text{real}(p_1), \ldots, \text{propositions formed from these using } \neg, \rightarrow, \wedge, \vee\}$$
$$\downarrow \quad \text{v plus truth-tables}$$
$$\{\textsf{T}, \textsf{F}\}$$

In classical propositional logic every atomic proposition is abstracted to just its truth-value. Nothing else matters. So if we have two models in which for each propositional variable p_i the one model assigns to p_i a proposition that is true iff the other model assigns one that is true, then exactly the same formal wffs are evaluated as true via their realizations. In that sense, the models are indistinguishable. The propositions in one may be about mathematics and in the other about raising sheep.

Some logicians take this as a reason to abstract a model to:

Type **II**
$$L(\neg, \rightarrow, \wedge, \vee, p_0, p_1, \ldots)$$
$$\downarrow \quad \text{v plus truth-tables}$$
$$\{\textsf{T}, \textsf{F}\}$$

Here v assigns truth-values to the variables $p_0, p_1, \ldots$ and is extended to all formal wffs by the classical truth-tables for $\neg, \rightarrow, \wedge, \vee$. They often name this abstracted version of a model by its valuation, v.

In doing this, the language and semantics of classical propositional logic have become formal and abstract. Simplification and abstraction are important because they allow us to use mathematics to establish general results about our logic. But these results will have significance only if we remember that models of type **II** come from models of type **I**.

In mathematizing classical propositional logic, certain collections are treated as sets within the mathematics of set-theory. Formal set-theory is rarely used; what is assumed is roughly what you can find in a mathematician's presentation, such as Felix Hausdorff's *Set Theory*.

So, for example, the collection of propositional variables, PV, is treated as a set. It is an infinite set, not just a way to generate further propositional variables as needed. The collection of well-formed formulas, Wffs, is treated as a completed infinite set. Looking only at models of type **II**, a valuation is a function $v: PV \rightarrow \{\textsf{T}, \textsf{F}\}$. This function is extended inductively to all wffs by the truth-tables, which are also functions. Then the mathematical logician makes a powerful generalization.

The Fully General Classical Abstraction of Propositional Logic Models
Any function $v: PV \to \{T, F\}$ which is extended to all wffs by the truth-tables is a model.

Thus, not only does it matter how a model of type **II** arises, the mathematician assumes that any mathematically possible one can arise. In particular, it's assumed that we can independently assign truth-values to all the variables in PV. All the definitions in Chapter 4 are then to be understood in terms of these models and assumptions.

These assumptions are not about logic but about how to do the mathematics of *metalogic*, the methods of reasoning we use to reason about logic. They apply to models of type **II**, not type **I**. The simplicity and generality we get by making these assumptions is justified so long as we don't arrive at any contradiction with our previous more fundamental assumptions and intuitions when we apply a metalogical result to actual propositions.

Someone who holds a platonist conception of logic and views propositions as abstract things would most likely take as basic facts about the world what I have called assumptions, idealizations, and generalizations. Words and sentences are understood as abstract objects that sometimes can be represented as inscriptions or utterances. They are not equivalences we impose on the phenomena of our experience because all of them, the platonist says, exist whether they are ever uttered or not. The formal and informal language and models are abstract objects complete in themselves, composed of an infinity of things. Though the distinction between a model of type **I** and **II** can be observed by a platonist, it does not matter whether a model is ever actually expressed or brought to our attention. It simply exists, and hence the Fully General Classical Abstraction is no abstraction but an observation about the world.

The Extension of a Predicate

Consider this model of classical predicate logic, where the predications are given their "obvious" values.

$$L(\neg, \to, \wedge, \vee, \forall, \exists; P_0, P_1, \ldots, c_0, c_1, \ldots)$$
$$\downarrow$$
$L(\neg, \to, \wedge, \vee, \forall, \exists;\ —$ is a dog $—$ is a cat, $—$ eats grass, $—$ is a wombat, $—$ is the father of ; Ralph, Dusty, Howie, Juney)

universe: all animals, living or toy
$$\downarrow$$
$$\{T, F\}$$

Recall the classical abstraction of predicates: the only properties of a predicate we consider are its arity and for each (sequence of) object(s) under consideration whether it applies to that (sequence of) object(s) or not. Thus, if we were to evaluate the predicates and names as we ordinarily do, and we take another model that differs from the one above only in that it realizes P_1^1 as "$—$ is a domestic feline $—$" and realizes P_0^2 as "$—$ is the male parent of $—$", then the differences between the two models would be only in the semi-formal language, since "is a domestic feline" and "is a cat" apply to the same things in the universe, and "is the father of" and "is the male parent of" are true of the same pairs of things from the universe. We could identify the two models since they have the same universe and validate the same formal wffs.

To come up with a way to talk about such identifications of models and to simplify what we need to pay attention to in presentations of models, we make the following definition.

The extension of a predicate The collection of sequences of the universe of which a predicate is true is the *extension* of the predicate *relative to* that universe.

We can extend this definition to all open wffs, not just atomic predicates by:

If A is an open wff with n distinct free variables, then the *extension* of A is the collection of (sequence of) n objects of which A is true.

Thus, relative to a universe of all objects with diameter greater than 3 cm on the moon, "— is a dog" and "— is a cat" have the same extension. Relative to all objects with diameter greater than 3 cm in my house, "— likes broccoli" and "— is a human" have the same extension. For the universe of all living things with the usual truth values for predications, the following wffs have the same extension:

$(-$ is a half brother to $-)\,(x, y)$

$(-$ is human$)\,(x) \wedge (-$ is human$)\,(y) \wedge (-$ is male$)\,(x) \wedge$
$\exists!z\,((-$ is a parent of $-)\,(z,x) \wedge (-$ is a parent of $-)\,(z,y)$

Collections of objects and sequences of objects are central to the discussion of the extensions of predicates. But if collections are to play a role in our abstraction of classical models, then, comparable to the classical abstraction of predicates, how we specify a collection cannot matter:

Identity Condition for Collections
$A = B$ iff for any object c, $c \in A$ iff $c \in B$

So relative to all living animals, the collection of objects x such that x is domesticated and x is a canine is the same as the collection of objects x such that x is a dog.

Given a model, we write:

U is the universe of a model viewed as a collection.

P_n^i is the realization of P_n^i
    P_n^i is the extension of P_n^i viewed as a collection of n-tuples.

c_i is the realization of c_i
    c_i is the object assigned to c_i in the model.

In the model above, the realization of P_0^2 is "— is the father of —", and the extension of "— is the father of —" is the collection of pairs of animals in which the first is the father of the second. The realization of c_0 is "Ralph", and the object assigned to the name "Ralph" is what's pictured on the back cover of this book.

Schematically, a model of type **I** is then:

I $L(\neg, \rightarrow, \wedge, \vee, \forall, \exists, =; P_n^i, \ldots, c_0, c_1, \ldots)$ where $i, n \geq 0$
$\quad\quad\quad\quad\quad\downarrow \quad\quad\quad\quad$ realizations of name and predicate symbols; universe

$\quad L(\neg, \rightarrow, \wedge, \vee, \forall, \exists, =; P_n^i, \ldots, c_0, c_1, \ldots)$ where $i, n \geq 0$
$\quad\quad\quad\quad\quad\downarrow \quad\quad\quad\quad$ assignments of references; valuations of atomic wffs

$\quad\quad < \mathsf{U}; P_0^1, P_1^1, \ldots, P_n^i, \ldots, c_0, c_1, \ldots >$
$\quad\quad\quad\quad\quad\downarrow \quad\quad\quad\quad$ evaluations of $\neg, \rightarrow, \wedge, \vee, \forall, \exists, =$

$\quad\quad\quad \{\mathsf{T}, \mathsf{F}\}$

The classical abstraction of predicates requires that we ignore any differences between two models that appear only at the second level (the semi-formal language). If two models have the same interpretation, they will validate exactly the same formal wffs. So we can make a *classical abstraction of the model*.

II $L(\neg, \rightarrow, \wedge, \vee, \forall, \exists, =; P_n^i, \ldots, c_0, c_1, \ldots)$ where $i, n \geq 0$
$\quad\quad\quad\quad\quad\downarrow \quad\quad\quad\quad$ interpretation

$\quad\quad < \mathsf{U}; P_0^1, P_1^1, \ldots, P_n^i, \ldots, c_0, c_1, \ldots >$
$\quad\quad\quad\quad\quad\downarrow \quad\quad\quad\quad$ assignments of references; valuations of atomic wffs;
$\quad\quad\quad\quad\quad\quad\quad\quad\quad\quad\quad\quad$ evaluations of $\neg, \rightarrow, \wedge, \vee, \forall, \exists, =$

$\quad\quad\quad \{\mathsf{T}, \mathsf{F}\}$

Atomic predications are evaluated by the following for assignments of reference σ.

The extensional evaluation of predicates and functions
$\quad \vee_\sigma(P_n^i(t_1, \ldots, t_n)) = \mathsf{T}$ iff the sequence $(\sigma(t_1), \ldots, \sigma(t_n))$ is in $\mathbf{P}_n^i$

Using these evaluations we have a model of type **II** for the formal language. I'll use $M, M_0, M_1, \ldots$ as metavariables for type **II** models also.

In a model of type **II** there are no longer any semi-formal propositions to assign truth-values to. Rather, a *formal wff is to be considered a proposition under the interpretation*, for example, $\forall x_1 \exists x_2 (P_1^2(x_1, x_2) \vee P_{17}^1(c_{36}))$ is a proposition. With the meaning given to the symbols of an open formal wff by a model of type **II**, a wff is *satisfied* or not by an assignment of references, and a closed formal wff is *true* or *false*.

Now compare:

$\quad M_3 = < \{3, 6, 8, 10\}; \{x : x \text{ is prime}\} >$

$\quad M_4 = < \{3, 6, 9, 15\}; \{x : x \text{ is even}\} >$

These have different universes and different predicates in extension, yet the models are the same in the sense that we can map the universe of M_3 to the universe of M_4 via the function φ given by $\varphi(3) = 3, \varphi(6) = 6, \varphi(8) = 9$, and $\varphi(10) = 15$. Then we can map assignments of references σ on M_3 to assignments of references σ^φ on M_4 via $\sigma^\varphi(x) = \varphi(\sigma(x))$, and that preserves truth in a model: $\vee_\sigma(A) = \mathsf{T}$ iff $\vee_{\sigma^\varphi}(A) = \mathsf{T}$.

Just as all that matters about a predicate in this view is its extension, all that matters about the universe of a model is that it is a collection and the objects in that collection satisfy or do not satisfy various predicates. Abstracted that far, two models whose universes can be put into a correspondence that preserves truth in the models are just notational variants of each other.

Mathematical Models of Classical Predicate Logic

Taking the extension of a predicate to be a set, we can further abstract the notion of a model of classical predicate logic. In set theory U^n is the set of all sequences (viewed as abstract objects) of n things from the universe U. Then a *classical mathematical model* of the language $L(\neg, \rightarrow, \wedge, \vee, \forall, \exists, =; P_i^n, c_i)$ is:

(M) $\quad M = \langle U; P_0^1, P_1^1, \ldots, P_i^n, \ldots, \mathsf{a}_0, \mathsf{a}_1, \ldots \rangle$

1. U is a nonempty set.
2. For each $i \geq 0$, $n \geq 1$, P_i^n is a set and $P_i^n \subseteq U^n$.
3. For each $i \geq 0$, $\mathsf{a}_i \in U$.

An *assignment of references* is any function σ: Terms$\rightarrow U$ satisfying:

For each i, for any assignments of references σ and τ, $\sigma(c_i) = \tau(c_i)$.

We notate $\sigma(c_i) = c_i$. The collection of assignments of references is complete because there are enough such functions according to the tenets of set-theory.

We define inductively "σ *satisfies* A" as we did previously, except that for atomic wffs we set:

$\vee_\sigma(t_1 \equiv t_2) = T$ iff $\sigma(t_1) = \sigma(t_2)$

$\vee_\sigma(P_i^n(t_1, \ldots, t_n))$ iff $(\sigma(t_1), \ldots, \sigma(t_n)) \in P_i^n$

The extensionality condition is satisfied because of the identity condition for collections (p. 187).

The Fully General Classical Abstraction of Predicate Logic Models

Any structure satisfying the conditions at (M) is a model of the formal language.

We no longer require that a classical mathematical model arises as an abstraction of a type **I** or even a type **II** model. Any structure will do to interpret the formal symbols. All connection between the formal symbols and reasoning in ordinary language is cut. The motive for making this abstraction is a belief that a set can exist independently of any possibility of its being the extension of a linguistic predicate.

The fully general classical abstraction of models allows us to treat logic, including models and the notion of truth, as mathematical subjects, ones about which we can reason without regard to their original motivation, paying attention to structure only. It is a powerful abstraction and is the basis of what is now called *mathematical logic*.

The mathematical logician sees this as just a clearer way to express the underlying structural aspects of classical predicate logic. But it is much more than that. To proceed this way, we must agree that there are abstract objects, namely sets, sequences, functions, and more. Previously, we could consider the universe of a model to be either objects in a (perhaps large) place or objects picked out by a linguistic predicate. Now the universe of a model is not physical at all: the set of all pigs in Denmark is an abstract object.

Moreover, we must assume that there are infinite completed collections, such as the collection of all terms and the collection of all wffs of the formal language.

Further, in this conception of classical predicate logic, only one notion of truth is allowed. For a given universe, a unary predicate is a set P, and it is *true of an object* a of the universe means that a∈ P. This is not so far from what we did before. But a binary predicate Q on the universe is not true of any objects of the universe; it is a collection of sequences, and it is true of a sequence (a, b), where a sequence is an abstract thing.

Paying attention to structure only can lead us to sever logic from reasoning. It is not clear that a notion of set divorced from that of the extension of a linguistic predicate can be used in defining models and still yield a good formalization of the notion of validity. The classical mathematical logician argues that the notion of valid inference in classical mathematical logic is, for formal schemes, the same as the formalization of validity in classical predicate logic.

By the completeness theorem (Appendix 4)

$\Gamma \vdash A$ iff $\Gamma \vDash A$ in classical predicate logic.

$\Gamma \vdash A$ iff $\Gamma \vDash A$ in classical mathematical predicate logic.

Hence, $\Gamma \vDash A$ in classical mathematical predicate logic iff $\Gamma \vDash A$ in classical predicate logic.

But to establish the completeness theorem nonconstructive, infinitistic principles are used. As noted in Appendix 4 (p. 179), for those who do not accept that there are completed infinities that we can survey all at one go, it is open whether if $\Gamma \vDash A$ in classical predicate logic then $\Gamma \vdash A$. It might be that with fewer models, more semantic inferences are valid in classical predicate logic than in classical mathematical logic.

Appendix 7 Aristotelian Logic

The Tradition	191
Why These Forms?	192
Subject and Predicate	193
Some Classifications of Categorical Propositions	194
Exercises	195
Truth and the Square of Opposition	195
Exercises	197
Syllogisms	198
Exercises	199
Evaluating Syllogisms for Validity	200
Exercises	202

The Tradition
Aristotle was the first to investigate whether inferences are valid by inspection of their form, confining his attention to inferences composed of categorical propositions.

Categorical propositions A *categorical proposition* is one that is or can be rewritten as an equivalent proposition that has one of the following *standard forms*:

All S are P.

Some S is (are) P.

No S is (are) P.

Some S is (are) not P.

For any proposition in one of these forms, the *term* (word or phrase) that replaces the letter S is called the *subject* of the proposition. The term that replaces the letter P is called the *predicate* of the proposition.

Example 1: *No police officers are thieves.*
Analysis This has the form: No S is P. The subject of it is "police officers", the predicate is "thieves".

Example 2: *Some cats are nice.*
Analysis This has the form: Some S are P. It has subject "cats" and predicate "nice".

Example 3: *Some dog is brown.*
Analysis This has the form: Some S is P. It has subject "dog" and predicate "brown".

Example 4: *Every dog is a canine.*
Analysis This can be rewritten as "All dogs are canines", which is categorical with form: All S are P. It has subject "dogs" and predicate "canines".

Example 5: *Some cats are not black.*
Analysis This has the form: Some S is not P. It has subject "cats" and predicate "black".

From Aristotle's time until the late 1800s, his work as supplemented by the work of others on the role of propositional connectives in reasoning, was the basis for most argument analysis. That tradition, called *Aristotelian logic*, became a subtle tool for the analysis of inferences in the Middle Ages, especially from about 1100 to 1400.

In the late 1500s, scholars became more interested in studying informal reasoning, inspired also by the work of Aristotle. They ignored the complexities of the formal logic of the medievals and were content with just the rules and forms of Aristotelian logic. That simplified tradition of Aristotelian logic, current since about 1600, is all I'll present here. It's worth studying because many writers from that time until today have used its terminology.

Aristotelian logic in its simplified methods is also worth studying because it presents a contrast with predicate logic. But it is only in the work of the medievals, which in the last 90 years has begun to be rediscovered, translated, and discussed, that the Aristotelian tradition can offer us anything in the way of a serious study of inferences and a guide to reasoning well.

Why These Forms?
The following propositions are in standard form:

>All dogs are mammals.
>No nurse is a doctor.
>Some newspaper is written in Arabic.
>Some snow is not white.

Many propositions we reason with aren't in one of these forms, yet seem similar to categorical ones.

(1)   All dogs bark.
>   No horse eats meat.
>   Some cats eat birds.
>   Some dogs don't chase cats.

To rewrite these as categorical propositions, we must have some idea of what semantic assumptions are used to justify our paying attention to just these forms.

Aristotle gave none. He simply took it as given that these forms are ubiquitous in reasoning. But we've seen throughout this text that without a semantic basis for deciding what forms we pay attention to and what our decisions about validity are based on, we are at a loss when we come to an example of an inference that we know isn't valid but seems to have the right form or when we come to an inference that has the wrong form but is valid. All we can point to is syntax, and that doesn't take us far.

In the Middle Ages, logicians settled on taking categorical propositions to be about things. Then they could rewrite the propositions at (1) as categorical:

>All dogs are things that bark
>No horse is a thing that eats meat
>Some cat is a thing that eats birds
>Some dog is not a thing that chases cats

Somewhat more colloquially, or at least avoiding the constant use of the phrase "thing that", we could rewrite these as:

>All dogs are barkers.

No horse is a meat eater.
Some cat is a bird eater.
Some dog is not a cat chaser.

If this is all there is to Aristotelian logic, it's just a fragment of what we developed with predicate logic.

But there are plenty of categorical propositions we use in inferences that are not about things. For example, the following inference is valid:

(2)   All snow is white.
      All that is white is visible.
      Therefore, all snow is visible.

To view this as about things, logicians of the Middle Ages read "all snow" as "all things that are snow" or "all bits of snow". We saw in Chapter 14.G that such a reading of mass terms is not compatible with an analysis of reasoning that requires the use of variables as temporary names. There are no variables here, but the same arguments strongly suggest that it is not coherent to reason about masses as made up of things.

The following inference also uses only categorical propositions and is valid.

(3)   Justice is loved by the gods.
      What is loved by the gods is desirable.
      Therefore, justice is desirable.

To convert this into propositions about things we'd have to replace the mass term "justice" with "what is just". Similarly, "Beauty is beautiful" is categorical and is not about a thing, unless you think that "beauty" and "justice" are names of things. That might be acceptable to a platonist, but many of the logicians who developed medieval logic were not platonists. They wanted to reduce talk of qualities and properties to talk of things.

What is interesting about Aristotelian logic is that it can be used to evaluate inferences such those at (2) and (3) that lie outside the scope of predicate logic precisely because they do not fit into the view of the world as made up of things. But to accept Aristotelian logic in those terms is to accept it without a clear semantic basis: the S and P in categorical forms can be words or phrases for collections of things, for individual things, for masses, for qualities, and more. There is a tension between (i) wanting to provide a semantic basis and hence narrowing the scope of what kinds of terms can be used in categorical propositions and (ii) wanting Aristotelian logic to have as broad a scope as possible. It is that tension that illuminates the issues of the relation of logic, language, and the world that we have considered with predicate logic.

I'll not resolve those issues here. I'll just point them out out and note that if Aristotelian logic is about things, it's a poor cousin of predicate logic, and if Aristotelian logic is about more of what is in the world, then it provides an interesting if unclear metaphysical contrast to predicate logic, with overlapping but not identical scope.

Subject and Predicate

Example 6: *All cows are white.*
Analysis This is in categorical form. The subject is "cows" and the predicate is "white". In

predicate logic we would say that the predicate in this example is "are white", and we would not talk about the subject, for subjects are parsed as some combination of term(s) and predicate(s).

There was disagreement among Aristotelians whether a predicate and a subject of a categorical proposition are linguistic, a disagreement that persisted into predicate logic. In this example, some would say that the subject is cows and the predicate is white or whiteness, whether those are collections, or abstract universals, or mental entities. Here I'll follow the view that subjects and predicates are or can be expressed or represented by linguistic phrases.

Example 7: *Some horse is a stallion.*
  Some horses are stallions.
Analysis Both of these are in categorical form. Aristotelians treat them as the same because they have the same subject and the same predicate. But to do so, they have to identify "a stallion" with "stallions".

Example 8: *Some cats eat birds.*
Analysis We said we could rewrite this as "Some cat is a bird eater", which is in categorical form. But to accept the rewrite we have to justify why we should treat "eat birds" and "a bird eater" as equivalent, for they certainly aren't the same linguistic phrase. From a nonlinguistic view of subjects and predicates, we would have to justify why those two phrases pick out or represent the same abstract object or mental entity.

Some Classifications of Categorical Propositions

Aristotelians classified propositions that are in standard form.

Universal and particular propositions

Propositions in the form "All S are P" and "No S is P" are *universal*.

Propositions in the form "Some S is P" and "Some S is not P" are *particular*.

Example 9: *No police officers are thieves.*
Analysis This is classified as a universal proposition because it is taken to be equivalent to "All police officers are not thieves". Generally, "No S is P" is classified as universal because it is considered equivalent to "All S are not P".

Example 10: *Socrates is mortal.*
Analysis To widen the scope of their logic, Aristotelians say that this is a categorical proposition and classify it as universal.

Categorical propositions with a singular subject

Propositions of the form "c is P" or "c is not P" where c is a name are in standard form and are classified as universal categorical propositions.

Example 11: *Ralph is not a cat.*
Analysis This is a universal categorical proposition.

Affirmative and negative propositions

Propositions of the form "All S are P" and "Some S is P" are called *affirmative*.

Propositions of the form "No S is P" and "Some S is not P" are called *negative*.

Example 12: All dogs are mammals.
Analysis This is a universal affirmative proposition.

Example 13: No dog is a feline.
Analysis This is a negative universal proposition.

Example 14: Some dogs are not friendly.
Analysis This is a negative particular proposition.

Quality and quantity

Whether a proposition is universal or particular denotes its *quantity*.
Whether a proposition is affirmative or negative denotes its *quality*.

Exercises
1. What is a categorical proposition?
2. What is a universal categorical proposition? Give two examples.
3. What is a particular categorical proposition? Give two examples.
4. What is an affirmative categorical proposition? Give two examples.
5. What is a negative categorical proposition? Give two examples.
6. What does the quantity of a categorical proposition designate?
7. What does the quality of a categorical proposition designate?

For each of the following, state whether it is categorical or can be rewritten as an equivalent categorical proposition. If categorical, identify its subject, predicate, quantity, and quality.

8. All dogs are carnivores.
9. Some cat is not a carnivore.
10. Tom is a basketball player.
11. No fire truck is painted green.
12. Donkeys eat meat.
13. There is at least one chimpanzee who can communicate by sign language.
14. Every border collie likes to chase sheep.
15. No one who knows critical thinking will ever starve.
16. All dogs bark or Spot is not a dog.
17. Heroin addicts cannot function in a nine-to-five job.
18. Some people who like pizza are vegetarians.
19. Not every canary can sing.
20. Socrates did not have a cat.
21. If Zoe does the dishes, then Dick will take Spot for a walk.
22. Waiters in Las Vegas make more money than lecturers at the university there.

Truth and the Square of Opposition

Example 15: Some dogs are brown.
Analysis This is true just in case there is at least one dog.

Example 16: Some snow is white.
Analysis If you think that by "snow" what is meant is all bits of snow, then this is true just in case there is at least one bit of snow that is white. However, if you view snow as a mass, then this is true just in case there is some of that mass which is white.

196   An Introduction to Formal Logic

Example 17:   *Some justice is merciful.*
Analysis Perhaps you could view justice as a collection of all instances of justice, though I don't know what an instance of justice would be. Perhaps, justice is a mass, though it hardly seems like snow which is a physical mass. All we can say here is that if you take the sentence to be a proposition, then you'll have to explain what you mean by "some justice".

Example 18:   *All dogs bark.*
Analysis It's not enough, as we know from our work in predicate logic, to say this is true just in case all dogs bark. Does that require that there is at least one dog?

Aristotelians assumed that universal propositions have *existential import*: for the proposition to be true, the subject term has to stand for something that exists.

Example 19:   *Some dog is white.*
Analysis For this to be true, there has to be a dog.

Example 20:   *Some dog is not white.*
Analysis For this to be true, there has to be a dog.

Example 21:   *Socrates is mortal.*
Analysis For this to be true, Socrates has to exist, for recall that this is classified as universal.

Example 22:   *All dodos are birds.*
Analysis This is false because there are no dodos.

Example 23:   *No dodos are birds.*
Analysis This is false because it's understood to be equivalent to "All dodos are not birds" and there are no dodos.

Aristotelians devised a classification of pairs of propositions according to how their truth-values are related.

Contradictories, contraries, and subcontraries

Two propositions are *contradictory* if it is not possible for them to have the same truth-value.

Two propositions are *contrary* if it is not possible for them both to be true.

Two propositions are *subcontrary* if it is not possible for them both to be false.

They used the following diagram to summarize how these terms apply to pairs of categorical propositions.

The Square of Opposition

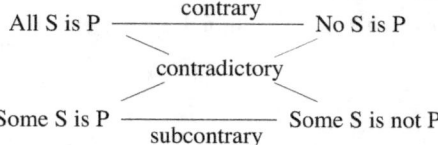

Where there is no line connecting the two forms, propositions of those forms are neither contradictories, nor contraries, nor subcontraries.

Example 24:   *All dogs are domesticated.*
       Some dogs are not domesticated.

Analysis According to the diagram these are contradictory. But if there were no dogs (horror of horrors), both would be false because they both have existential import, so they wouldn't be contradictory. Thus, they are contradictory only if "all" does not have existential import.

Example 25: *Some teacher is kind.*
   No teacher is kind.
Analysis According to the diagram, these are contradictories. But if there are no teachers, both would be false. So they are contradictory only if "all" does not have existential import.

Example 26: *All cats are nice.*
   No cat is nice.
Analysis According to the diagram, these are contraries. They can't both be true, for if either is true, there are cats. But they both could be false.

Example 27: *Some diamonds are precious stones.*
   Some diamonds are not precious stones.
Analysis According to the diagram, these are subcontraries, which is correct if there are diamonds. But if there are no diamonds, then both are false and these aren't subcontrary.

Despite being memorized by generations of students over hundreds of years, the diagram doesn't work. The relationships vary according to whether "all" and "some" have existential import.

Exercises
1. a. What does it mean to say that two propositions are contradictory?
   b. What does it mean to say that two propositions are contrary?
   c. What does it mean to say that two propositions are subcontrary?

   For each pair of propositions below, state which of the terms apply:
   contradictories
   contraries
   subcontraries
   none

2. All dogs bark.
   Some dogs do not bark.
3. No Russians are communists.
   All Russians are communists.
4. Maria is a widow.
   Maria was never married.
5. No animals with horns are carnivores.
   Some animals with horns are carnivores.
6. All uranium isotopes are highly unstable substances.
   Some uranium isotopes are highly unstable substances.
7. Some assassinations are morally justifiable.
   Some assassinations are not morally justifiable.
8. Dick and Tom are friends.
   Dick and Tom can't stand to be in the same room together.
9. Not even one zebra can be trained to jump through fire.
   Every zebra can be trained to jump through fire.
10. Homeless people don't like to sleep on the street.
    Some homeless people don't like to sleep on the street.

Syllogisms

Aristotelians were principally concerned with inferences that contain only categorical propositions. They reduced the classification of those as valid or invalid to an analysis of inferences of a particular kind.

Categorical syllogisms A *categorical syllogism* is an inference composed of three categorical propositions: two premises and a conclusion. Exactly three terms are used as subject or predicate in those propositions, and each term appears in exactly two of the propositions.

Example 28: *Some dogs are pets.*
  No dogs are feline.
  Therefore, *some pets are not feline.*

Analysis This is a categorical syllogism. The terms in it are "dogs", "feline", and "pets". Each appears in exactly two of the propositions. It has the form:

  Some S are P.
  No S are Q.
  Therefore, some S are not Q.

Is this valid? Are all other inferences of this form valid?

Aristotelians began their analysis by identifying the roles that the predicates and subjects play in syllogisms.

Major, minor, and middle terms of a categorical syllogism

  major term   the predicate of the conclusion
  minor term   the subject of the conclusion
  middle term the term that appears in both premises
  major premise the premise that contains the major term
  minor premise the premise that contains the minor term

A syllogism is in *standard form* if all the propositions in it are in standard form, and the major premise comes first, then the minor premise, then the conclusion.

Example 29: *No police officers are thieves.*
  Some thieves are sent to prison.
  Therefore, *no police officers are sent to prison.*

Analysis This is a categorical syllogism.

  The major term is "sent to prison".
  The minor term is "police officers".
  The middle term is "thieves".
  The major premise is "Some thieves are sent to prison".
  The minor premise is "No police officers are thieves".

It is not in standard form. The standard form for it is:

  Some thieves are sent to prison.
  No police officers are thieves.
  Therefore, no police officers are sent to prison.

Aristotelians used abbreviations for the kinds of categorical propositions in order to facilitate memorizing which forms have only valid instances.

A All S are P.
E No S is P.
I Some S is P.
O Some S is not P.

Example 30: *No dog is a cat.*
All cats are mammals.
Therefore, *no dog is a mammal.*
Analysis This has the form EAE. It is not valid.

Example 31: *All dogs are canines.*
All canines are mammals.
Therefore, *all dogs are mammals.*
Analysis This has the form AAA. It is valid.

Aristotelians showed that the question of which forms have only valid instances could be reduced to classifying just a few forms. Each of those could be classified by either producing an invalid inference of that form or arguing that each instance had to be valid. It was only in the late 1700s that a routine method was given for determining whether a categorical syllogism is valid, which we'll see in the next section.

Exercises
1. What is a categorical syllogism?
2. Can a syllogism be valid with two categorical premises and a categorical conclusion which do not have exactly three terms?

For each of the following, determine whether it is or can be rewritten as a categorical syllogism. If it can be, then:
   i.   List the major term and the minor term.
   ii. State whether it is in standard form, and if it is not, put it into standard form.
   iii. Give the form of it using the AEIO abbreviations.

3. All students at this school pay tuition. Some people who pay tuition at this school will fail. So some students at this school will fail.
4. There aren't any wasps that will not sting. Some bumblebees will not sting. So some bumblebees aren't wasps.
5. No pacifist will fight in a war. Arf is a pacifist. So Arf will not fight in a war.
6. Badly managed businesses are unprofitable. No oyster cultivating business in North Carolina is badly managed. So some oyster cultivating business in North Carolina is profitable.
7. Nothing that's smarter than a dog will cough up hair balls. Cats cough up hair balls. So cats are not smarter than dogs.
8. Dick will not visit Tom tonight if Zoe cooks dinner. Zoe didn't cook dinner. So Dick visited Tom tonight.
9. Police chiefs who interfere with the arrest of city officials are always fired. People who are fired collect unemployment. So some police chiefs who interfere with the arrest of city officials collect unemployment.

10. Some temporary employment agencies do not give employee benefits. All employees of Zee Zee Frap's restaurant get employee benefits. So no employee of Zee Zee Frap's is hired through a temporary employment agency.
11. What is beautiful is good. What is good is loved by the gods. Beautiful people are beautiful. So beautiful people are loved by the gods.
12. Steak is meat. What is meat is loved by the dogs. So steak is loved by the dogs.

Evaluating Syllogisms for Validity

Example 32: *All good teachers give fair exams.*
Prof. Zzzyzzx gives fair exams.
So Prof. Zzzyzzx is a good teacher.
Analysis This isn't in the form of a categorical syllogism, but with the practice you've had with the last set of exercises you can rewrite it. I'll deal directly with the example.

We can show that the example isn't valid by giving an example of a way the world could be in which the premises are true and conclusion false: Prof. Zzzyzzx could be a terrible teacher who gives fair exams from an instructor's manual. That suffices, but it requires us to think of a possibility. We can make the analysis routine by drawing a diagram.

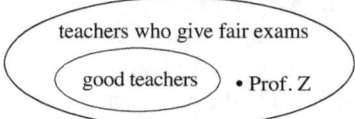

The outer circle represents all teachers who give fair exams. The first premise says that all good teachers give fair exams, so the circle representing the good teachers is within the circle of teachers who give fair exams. The second premise says that Prof. Z gives fair exams, so we have to put a dot for him within the big circle for good teachers. But that's the only restriction we have, so we can put it outside the circle for good teachers. That is, Prof. Z need not be a good teacher. So the inference isn't valid.

Checking for validity with diagrams

- A collection is represented by an enclosed area.
- If one area is entirely within another, then everything in the one collection is also in the other.
- If one area overlaps another, then there is something that is common to both collections.
- An "a" or a dot in an area marks that a particular object is in that collection
- Draw the areas to represent the premises as true while trying to represent the conclusion as false. If you can, then the argument is invalid. If there's no way to represent the premises as true and the conclusion as false, the argument is valid.

Example 33: *All dogs bark.*
Everything that barks is a mammal.
Therefore, *All dogs are mammals.*
Analysis Again I'll work directly with the example and leave to you to rewrite it as a categorical syllogism in standard form.

We first draw the diagram to represent the premises as true.

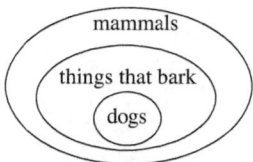

The dogs area is completely within the things that bark area. The things that bark area is completely within the mammals area. So the dogs area ends up being inside the mammals area. So if we represent the premises as true, we are forced to represent the conclusion as true. The argument is valid.

Example 34:   *Some kangaroos are tame.*
    Some creatures that are tame live in New Zealand.
    Therefore, *Some kangaroos live in New Zealand.*

Analysis We draw a diagram to represent the premises as true and conclusion false:

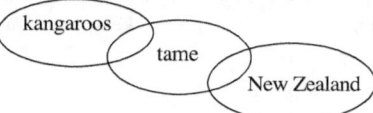

For the first premise, the kangaroos area must overlap the area for tame things. For the second premise, the area representing tame things has to overlap the area for things that live in New Zealand. We can draw that without making the areas for kangaroos overlap that of things that live in New Zealand. So it doesn't follow that some kangaroos live in New Zealand. The argument is invalid.

Example 35:   *All dogs bark.*
    No professor is a dog.
    Therefore, *No professor barks.*

Analysis For the first premise we have to draw the area for dogs within the area for things that bark. For the second premise we have to draw the area for professors entirely outside the area for dogs. Here are three ways we can do that.

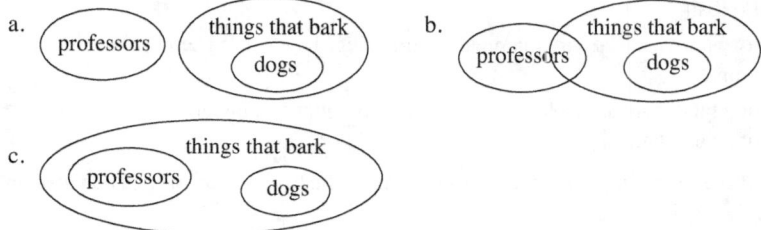

These (schematically) represent all the ways the premises could be true. Yet in both (a) and (b) the conclusion is represented as false. It's possible for there to be a professor who barks, even though he (she?) isn't a dog. Arf, arf.

Example 36:   *Justice is loved by the gods.*
    What is loved by the gods is desirable.
    Therefore, *Justice is desirable.*

Analysis This is valid. We can show that with a diagram. But to do so commits us to viewing justice as being a collection. A collection of what? Just acts? Instances of justice? Alternatively, we could view "justice" as a name for a thing, so that the first premise and conclusion would be singular. Either seems at best odd. Yet Aristotelians were happy to use the method for inferences like these, too.

Exercises

Which of the following inferences are valid? Justify your answer.
You need not rewrite each as an Aristotelian syllogism in standard form.

1. Not every student attends lectures. Lee is a student. So Lee doesn't attend lectures.
2. No professor subscribes to *Rolling Stone* magazine. Maria is not a professor. So Maria subscribes to *Rolling Stone* magazine.
3. No professor subscribes to *Rolling Stone* magazine. Lou subscribes to *Rolling Stone* magazine. So Lou is not a professor.
4. Everyone who is anxious to learn works hard. Dr. E's students work hard. So Dr. E's students are anxious to learn.
5. All CEOs of Fortune 500 companies earn more than $400,000. Ralph earns more than $400,000. So Ralph is a CEO of a Fortune 500 company.
6. All students who are serious take critical thinking in their freshman year. No one who smokes marijuana every week is a serious student. So no one who smokes marijuana every week takes critical thinking his or her freshman year.
7. No student who cheats is honest. Some dishonest people are found out. So some students who cheat are found out.
8. Only ducks quack. George is a duck. So George quacks.
9. Everyone who likes ducks likes quackers. Dick likes ducks. Dick likes cheese. So Dick likes cheese and quackers.
10. No dogcatcher is kind. Anyone who is kind loves dogs. So no dogcatcher loves dogs.
11. Some things that grunt are hogs. Some hogs are good to eat. So some things that grunt are good to eat.
12. Dogs are loyal. Dogs are friendly. Anything that is friendly and loyal makes a great pet. Hence, dogs make great pets.
13. All dogs chase cats. All cats chase songbirds. So all dogs chase songbirds.
14. Some paraplegics can't play basketball. Belinda is a paraplegic. So Belinda can't play basketball.
15. Dogs are mammals. Cats are mammals. Some dogs hate cats. Therefore, some dogs hate mammals.
16. Everything made with chocolate is delicious. No liquor is delicious. So no liquor is made with chocolate.

Which of the following inference forms are valid? Justify your answer. Then give an argument of that form.

17. All S are P.
    No Q is S.
    So some Q aren't P.

18. All S are P.
    a is S.
    So *a* is P.

19. Some S are P.
    All P are Q.
    So some S are Q.

20. Only S are P.
    a is S.
    So *a* is P.

21. Some S aren't P.
    So no P are S.

22. All S are P.
    No Q is P.
    So no Q is S.

Index of Symbols

iff 9

$\neg, \to, \wedge, \vee$ 14

$p_0, p_1, \ldots$ 15

$A, B, C, A_0, A_1, A_2, \ldots$ 16, 20, 61

$L(p_0, p_1, \ldots, \neg, \to, \wedge, \vee)$ 16

p, q, r, s 16

$\top, \mathsf{F}$ 21

v 25, 86

real( ) 25

$\mathsf{M}, \mathsf{M}_0, \mathsf{M}_1, \ldots$ 26, 86, 189–190

Γ, Σ, Δ 29

$$\frac{p_7 \vee p_{10} \quad \neg p_{10}}{p_7} \quad 29$$

$\vDash A$ 26
$A \vDash B$ 29
$\Gamma \vDash A$ 29
$\Gamma \nvDash A$ 29

PC 32

$\leftrightarrow$ 38

$\vdash$ 46
$\Sigma \vdash A$ 46
$\Sigma \nvdash A$ 46

— (blank) 56

$x_1, x_2, x_3, \ldots$ 57

x, y, z 57, 108

$x, y, z, w, y_0, y_1, \ldots$ 61

$\forall$ 58
$\exists$ 58

$i, j, k, m, n, i_0, i_1, \ldots$ 61

$t, u, v, t_0, t_1, \ldots$ 61

$P_0^1, P_0^2, P_0^3, \ldots, P_1^1, P_1^2, P_1^3, \ldots$ 61

$P_0, P_1, \ldots$ 61

$c_0, c_1, \ldots$ 61

P, Q 61

$L(\neg, \to, \wedge, \vee, \forall, \exists, P_0, P_1, \ldots, c_0, c_1, \ldots)$ 63

$\forall \ldots$ 65

$\sigma, \sigma_0, \sigma_1, \ldots, \tau, \tau_0, \tau_1, \ldots$ 79

$\tau \sim_x \sigma$ 81

$\sigma \vDash$ 85

v_σ 81

$\mathsf{v}_\sigma \vDash$ 81

$C(A)$ 91

$A(x)$ 100

$A(t/x)$ 100

has_p 128

$\equiv$ 139

$=$ 139

$\not\equiv$ 140

$\exists_{\geq n}$ 145

$\exists_{\leq n}$ 145

$\exists !_n$ 146

$\exists !$ 146

$\cap \ \cup \ \subseteq \ \in$ 164

PV, Wffs 185

A, E, I, O 199

S, P 191

Index of Examples

- Only numbered examples with analysis are indexed.
- No purely formal example is indexed.
- Alphabetical order: blanks; parentheses; spaces; commas; letters; numerals; $\wedge, \vee, \neg, \rightarrow, \forall, \exists$.

A cat is mewing. 110
A cat is never your friend. 129
A dog is a dog. 110
A horse will eat an apple. 127
A horse will eat an apple, but not a dog. 127
A horse will eat an apple, but not an orange. 127
A mammal is an ungulate if it has hoofs. 9
All but 2 bears in the London Zoo are brown. There are 47 bears in the London Zoo.
    Therefore, there are 45 brown bears in the London Zoo. 152
All cats are nice.
    No cat is nice. 197
All cows are white. 193
All dodos are birds. 196
All dogs and all cats hate each other. 93
All dogs are canines. All canines are mammals. Therefore, all dogs are mammals. 199
All dogs are domesticated.
    Some dogs are not domesticated. 196
All dogs are mammals. 195
All dogs bark. 57, 82, 196
All dogs bark and all cats meow. 110
All dogs bark, and Humberto barks. So Humberto is a dog. 4
All dogs bark, and Humberto is a dog. So Humberto barks. 4
All dogs bark. Birta is a dog. Therefore, Birta barks. 121
All dogs bark. Everything that barks is a mammal. Therefore, All dogs are mammals. 200
All dogs bark. No professor is a dog. Therefore, No professor barks. 201
All dogs bark. Ralph is a dog. So Ralph barks. 36, 51, 54, 108
All dogs bark. Therefore, some dog barks. 123
All good teachers give fair exams. Prof. Zzzyzzx gives fair exams.
    So Prof. Zzzyzzx is a good teacher. 51, 200
All polar bears are white. 109
All polar bears in Antarctica can swim. 82
All sakes bark. Ralph is a sake. Therefore, Ralph barks. 53
All snow is white. All that is white is visible. Therefore, all snow is visible. 193
Almost all dogs bark. 147
At the end of this term, some of my students will get an A. 83

Bad dog, Spot, bad dog! 1
Barking is in a dog's nature. 131
Bon Bon is a small donkey. Therefore, Bon Bon is small. 117

Cows are not white. 113

Index of Examples

Dick and Tom lifted the table. 130
Dick doesn't have any cats. 128
Dick has a cold. Therefore, Dick has something. 129
Dick has a dog. Therefore, Dick has something. 128
Dick is a bachelor. Therefore, Dick is a man. 115
Dick is upset. 2
Dick or Zoe will go to the grocery to get eggs. 7
Dick will go into the army if and only if there is a draft. 9
Dick will go into the army only if there is a draft. 9
Dogs have eaten apples. 128

Either a Democrat won the election or a Republican won the election. 7
Either Dick picked up Zoe at the market, or Zoe went to see Suzy.
    Zoe didn't go to see Suzy. So Dick picked up Zoe at the market. 7
Either Ralph is a dog or Howie is a duck. No way is Howie a duck.
    Therefore, Ralph is a dog. 29
Every counting number is even or odd. 40, 123
Every dog doesn't meow. 113
Every dog hates every cat. 111
Every dog hates some cat. 112
Every dog is a canine. 190
Every dog which yelps also barks. 119
Every dog which yelps or whimpers also barks. 119
Every honest person is not a politician. 150
Every person who owns a puppy is not depressed. 119
Every puppy barks. 121
Everyone has seen Marilyn Monroe on TV. 124
Everyone has two parents. 152
Everything is a dog. There's not something that isn't a dog. 94
Everything is bigger than something. 148
Everything is somewhere. 130

George is a duck and Ralph is a dog. Therefore Ralph is a dog. 36
Get me a beer, Zoe. 1
Gold is more valuable than iron. 130

(— has the same blood type as —) (x, x) 72
(— has the same blood type as —) (x, y) 72
Horses eat apples. 126
Humberto got out of jail or he didn't get parole. Humberto did get parole.
    So Humberto got out of jail. 8

I am 1.80 m tall. 3
I wish I could get a job. 1
If anything is a dog, then Ralph is a dog. 124
If Dick goes to the basketball game, then either he got a free ticket
    or he borrowed money for one. 9
 If Dick takes Spot for a walk, then Zoe will cook dinner. And if Zoe cooks dinner, then
    Dick will do the dishes. So if Dick takes Spot for a walk, then he'll do the dishes.
    But Dick did take Spot for a walk. So he must have done the dishes. 11

If Donald Trump is honorable but Donald Trump lied, then Donald Trump is honorable. 37
If George is a duck then Ralph is a dog and Dusty is a horse. 14
If it's the day for the garbageman, then Dick will wake up. It's not the day for the garbageman. So Dick didn't wake up. 11
If Maria doesn't call Manuel, then Manuel will miss his class. Maria did call Manuel. So Manuel didn't miss his class. 11
If not both Ralph is a dog and Ralph barks, then if Ralph barks, he is a dog. 27
If Ralph is a dog but George is a duck, then George is a duck. 37
If Ralph is a dog or Howie is a cat, and Ralph is not a dog, then Howie is a cat. 26
If Ralph is a dog, then anything is a dog. 124
If Ralph is a dog, then he's a greyhound. So if Ralph is not a greyhound, he's not a dog. 38
If Ralph is a dog, then Ralph barks. Ralph barks. So Ralph is a dog. 38
If Ralph is a dog, then Ralph barks or Ralph howls. 14
If something weighs over 30 lbs., it cannot be sent by parcel post. 124
If Spot barks, then Dick will wake up. Dick didn't wake up. So Spot didn't bark. 12
If Spot barks, then Dick will wake up. Dick woke up. So Spot barked. 11
If Spot barks, then Dick will wake up. Spot barked. So Dick woke up. 10
If Spot barks, then Suzy's cat will run away. 9
If Spot got out of the yard, he was chasing a squirrel. 9
If Spot ran away, then the gate was left open. 8
If Suzy studies hard, then Suzy will pass the exam. 7
If the moon is made of green cheese, then 2 + 2 = 4. 40
I'll never talk to you again if you don't apologize. 8
I'm thinking of majoring in biology. That means you'll take summer school. 12
Inflation will be at least 3% this year. 8
(— is a dog) (Ralph) 83
(— is a dog) (Ralph) Therefore, $\exists x$ (— is a dog) (x) 88
((— is a dog) (Ralph) → (— barks Ralph)) →
    (¬ (— barks Ralph) → ¬ (— is a dog) (Ralph)) 90
(— is a dog) (z) ∧ (— is a cat) (x) → ¬(— is the father of —) (x, y) 99
(— is a parent of —) (x, y) 58
(— is a wombat) (Ralph) 81
(— is as tall as —) (x, y) 100
It's raining. 3, 131

Juney is barking loudly. Therefore, Juney is barking. 118
Justice is loved by the gods. What is loved by the gods is desirable.
    Therefore, Justice is desirable. 193, 201

Lee will pass his exam because he studied so hard. 7
(— loves —) (x, Ralph) 71
Loving someone means you never throw dishes at him. 8

Maria got the van or Manuel won't go to school. 8
Marilyn Monroe had all the qualities of a great actress. 131
Marilyn Monroe is a movie star. Marilyn Monroe is Norma Jean Baker.
    Therefore, Norma Jean Baker is a movie star. 148
Marilyn Monroe married Arthur Miller. Marilyn Monroe was Norma Jean Baker.
    Therefore, Norma Jean Baker married Arthur Miller. 148

Marilyn Monroe ≡ Norma Jean Baker Therefore,
    (— was an actress) (Marilyn Monroe) → (— was an actress) (Norma Jean Baker) 141
Mt. Everest has been climbed. Therefore, Someone has climbed Mt. Everest. 126

No cat can swim. 1
No dodos are birds. 196
No dog is a cat. All cats are mammals. Therefore, no dog is a mammal. 199
No dog is a feline. 195
No dog is a parent of a cat. 147
No dog meows 146
No dog meows. Juney is a dog. Therefore, Juney does not meow. 149
No police officers are thieves. 191, 194
No police officers are thieves. Some thieves are sent to prison.
    Therefore, no police officers are sent to prison. 198
Not only dogs bark. 151
Nothing both barks and meows. 147
Nothing is as dumb as a sheep. 149
Nothing is bigger than everything. 149
Nothing is friendlier than a dog. 149
Nothing is friendlier than Birta. 149

Only dogs bark. 150
Only dogs hate cats. 150
Only Ralph barks. 150

Pegasus is a horse. Therefore, there exists a horse. 123
Peter is in a car. Therefore, there exists a car. 53
Peter is in a hurry. Therefore, there exists a hurry. 53
Prince Albert married Queen Victoria. Queen Victoria is a head of state.
    Therefore, Prince Albert married a head of state. 129
Prince Albert married Queen Victoria. Queen Victoria is a widow.
    Therefore, Prince Albert married a widow. 130

Ralph is a dog and George is a duck and Howie is a cat. 37
Ralph is a dog and Juney is a dog. Therefore, there are two dogs. 152
Ralph is a dog and Ralph is not a dog. 36
Ralph is a dog but Howie is a cat. 36
Ralph is a dog if he's not a puppet. 38
Ralph is a dog only if he is not a puppet. 38
Ralph is a dog or Dusty is a horse and Howie is a cat. Therefore, Howie is a cat. 37
Ralph is a dog or Ralph is not a dog. 36
Ralph is a dog puppet. So Ralph is a dog. 36
Ralph is a dog $\land \neg$ (Ralph is a dog) 28
Ralph is a dog $\lor \neg$ (Ralph is a dog) 26
Ralph is a puppet; he is not a cat. 39
Ralph is a purple dog. Therefore, Ralph is purple. 117
Ralph is not a cat. 194
Ralph is not a dog because he's a puppet. 39
Ralph is the father of Birta. 151
Rose rose and picked a rose. 3
Running is fun. 121

Snow is white. All that's white is not black. Therefore, snow is not black. 120
Snow is white. What's white reflects light. So snow reflects light. 51
Socrates is mortal. 194, 196
Some cat is feral. 58, 109
Some cats are feral. 152
Some cats are nice. 191
Some cats are not black. 191
Some cats eat birds. 194
Some diamonds are precious stones.
    Some diamonds are not precious stones. 197
Some dog and some cat hate each other. 93
Some dog barks. 83
Some dog barks at all women and men. 126
Some dog hates all cats. 112
Some dog is brown. 191
Some dog is not white. 196
Some dog is white. 196
Some dogs are brown. 195
Some dogs are not friendly. 195
Some dogs are pets. No dogs are feline. Therefore, some pets are not feline. 198
Some dogs like cats. Some cats like dogs. So some dogs and cats like each other. 51
Some dogs like some cats. 152
Some horse is a stallion.
    Some horses are stallions. 194
Some kangaroos are tame. Some creatures that are tame live in New Zealand.
    Therefore, some kangaroos live in New Zealand. 201
Some justice is merciful. 196
Some snow is white. 195
Some teacher is kind.
    No teacher is kind. 197
Someone has climbed Mt. Everest. Therefore, Mt. Everest has been climbed. 125
Something is a dog.
    Not everything isn't a dog. 94
Something is better than nothing. 149
Something is bigger than everything. 148
Spot es un perro. 2
Spot is a dog.
    Spot is not a dog. He's a cat. 2
Spot is a dog and Puff is a cat. 38
Spot is a dog but Puff is not a dog. 37
Spot is barking 8
Spot is chewing your shoe. 1
Spot is out of the yard. 1
Spot is owned by Dick. Therefore, Dick owns something. 125
Spot is owned. Therefore, something owns Spot. 125
Stanisław Krajewski thinks that Marilyn Monroe was blonde. 72
Suzy took off her clothes and went to bed. 40

There are at least three dogs. 144
There are at least two dogs. 144
There are at least 7 but fewer than 42 wombats. 146
There are at least 47,813,201 stars. 145
There are at most two nice cats. 145
There are exactly 47 dogs that do not bark. 146
There are infinitely many stars. 145
There are more dogs than cats. 153
There are 2 prime numbers that differ by 2. 153
There is a nice cat. 124
There is a dog that meows and a cat that barks. 111
There is a woman who is a sister of a man who hates all cats. 126
There is no honest person who is a politician. 150
There is no tallest person. 150
Three faces of a die are even numbered. Three faces of a die are not even numbered.
    So cats bark. 39
Through any two points there is a line. 116
Through any two points there is exactly one line. 153
Tom or Suzy will pick up Manuel for class today. 8

Wanda wonders whether Ralph is a puppet. And Ralph is a puppet. 40
"(— was the teacher of —)" is true of Socrates and Plato. 72
What time does the movie start? 1
Whatever Ralph did yesterday, he is doing again today. 130
Whatever swims has gills. 130

Zeke disrespected Zoe. 116
Zélia is unmarried. 114
Zoe is a beautiful woman. 118

2/3 of all cats have fleas. 147
2 + 2 = 5   2
7 is less than 12. 117

$\neg\,((-\text{ is a dog})(x) \wedge \neg(-\text{ is a dog})(x))$
    $\rightarrow \neg\,((-\text{ is a dog})(x) \wedge \neg(-\text{ is a dog})(x))$ 90
$\forall x\,(-\text{ is a dog})(x)$ 98
"$\forall x\,((-\text{ is a dog})(x)$" is true   83
$\forall x\,(-\text{ is a dog})(x)$
    Therefore, $(-\text{ is a dog})(\text{Ralph})$ 88
"$\forall x\,((-\text{ is a dog})(x) \vee \exists x\,(-\text{ eats grass})(x))$" is true   85
$\forall x\,((-\text{ is a dog})(x) \rightarrow (-\text{ barks})(x)$
    $(-\text{ is a dog})(\text{Ralph})$
        Therefore, $(-\text{ barks})(\text{Ralph})$ 88
$\forall x\,(-\text{ is a dog})(x) \rightarrow (-\text{ is a dog})(\text{Ralph})$ 88, 90, 98
$\forall x\,(-\text{ is a dog})(x) \rightarrow \exists y\,(-\text{ barks})(y)$ 90
$\forall x\,((-\text{ is a dog})(x) \rightarrow \exists y\,(-\text{ is the father of }-)(y,x))$ 98
$\forall x\,((-\text{ is a man})(x) \vee (-\text{ is a woman})(x))$ 98

$\forall x$ (— is a man) (x) ∨ $\forall x$ (— is a woman) (x) 98
$\forall x$ (— is taller than —) (x, y) 99
"$\forall x \, \forall y$ (— is a cousin of —) (x, y)" is true   84
$\forall x \, \forall y$ (((— is a dog) (x) ∨ ¬ (— is a dog) (x)) ∨ (— is a cat) (y))   89
$\forall x \, \forall y$ [x ≡ y → ((— is a dog) (x) → (— is a dog) (y))]   141
$\forall x \, \exists y$ (— is a father of —) (y, x)
  $\exists y \, \forall x$ (— is a father of —) (y, x) 93
"$\forall x \, \exists y$ (— is the father of —) (y, x)" is true   84
"$\exists x$ (— is a dog) (x)" is true   83
$\exists x$ ((— is a dog) (x) ∧ (— barks) (x)) → ¬ $\forall y$ (— meows) (y) 91
$\exists x$ ((— is a dog) (x) ∨ ¬ (— is a dog) (x))   88
$\exists x$ (— is a dog) (x) → ¬ (— is a cat) (Ralph)   88
$\exists x$ ((— is a parent of —) (x, y))   59
$\exists x \, \exists y$ (¬ (— lives in the same house as —) (x, y))   98

Index

• Examples are indexed
  in the Index of Examples.

absolute adjective, 118
abstract object(s), 5–6, 73–74, 168,
  186, 190, 194.
  See also platonist.
adjectives, 117–119
  absolute, 118
  relative, 117
adverbs, 118–119, 134
affirmative categorical proposition, 194
affirming the consequent, 10
agreements, 116, 117, 165–166, 167–169
"all", 57, 82
  as conjunction, 155
  existential assumption?, 82, 109, 109,
    123, 196
  related to "some", 94, 103, 123
  See also relative quantification.
"almost all", 147
alternatives, 7
analysis vs. formalization, 115–116, 117, 123
"and", 8, 14,
  as "and then", 39
  truth-table for, 22
antecedent, 8, 14
anti-tautology, 28, 36, 40, 108
"anyone", "anybody", 124, 134
argument, 4
Aristotelian logic, 191–201
Aristotle, 191–192
arithmetic, 104, 152–153.
  See also geometry; *Index of Examples*.
arity, 56
assignments of references, 79–81
  atemporal —, 129
  completeness of collection of —, 80
  — for every variable, 80
  mathematical —, 189
atomic predicate?, 92
atomic proposition, 20, 57, 58
  connect predicates to named things, 70–71
  truth-value is given, 25, 81, 158–159

atomic wff,
  predicate logic, 62
  predicate logic with equality, 141
  propositional logic, 16,
attention, paying, 13, 88, 131, 158, 165–169
axiom, 43
  axiom scheme, 43, 46
  meaning —, 115–117
axiom system, 43
  complete —, 48
  predicate logic, 102–103
  predicate logic with equality, 141
  propositional logic, 43
  sound —, 47

basis of induction, 162
"because", 7, 39
biconditional, 38
binary predicate, 56
blank, 56
bound variable, 62
"but", 36–37
"by PC", 90, 175

Carnielli, Walter, 45, 96, 104, 173
categorematic vocabulary, 65
  formalizing and, 115–117
  meaning of —, 115–117, 131
  See also analysis vs. formalization;
    syncategorematic vocabulary.
categorical proposition, 191
categorical syllogism, 198
certainty, 166
Classical Abstraction, 21
  Fully General, 189
classical abstraction of a model, 188
Classical Abstraction of Names, 69
classical evaluation of $\forall, \exists$, 85
classical predicate logic, 89
  axiom system for, 102–103,
    completeness of —, 103, 178
    proof, 174–179
  sound, 174
  model for —. *See* model(s).

classical predicate logic with equality, 141
   axiom system for, 141
      completeness proof, 180–181
classical propositional logic, 32
   axiom system for, 43
      proof of completeness, 170–173
   model for —, 25–26
classical truth-tables, 21–23, 40
Clavius' law, 33
closed wff(s), 62
   propositions are —, 65
closure of a wff, 65
commands, 1
commas 14, 31, 56, 75
   in formal language, 14
common nouns
   — into predicates, 113
   reference and, 166
communication, 166
compactness
   predicate logic, 178
   propositional logic, 172
comparisons, formalizing, 148–150
complete theory, 48
completeness of axiom system. *See name of the system.*
completeness of collection of assignments of references, 80
compositionality,
   predicate logic, 81–82
   propositional logic, 21
compound predicate?, 92
compound proposition, 7
compound wff,
   in predicate logic, 62
   in propositional logic, 16
concepts, 74
conclusion of an inference, 4
conditional, 8, 14
   truth-table for, 22–23, 40, 57
conjunct, 14
conjunction, 14
   associativity, 33
      convention on —, 37
   commutativity, 33
   simplification law, 33
   tautology principle, 33
   temporal, 40

conjunction of terms, 130
connectives, 13–14
   with open formulas, 84
consequence, semantic, 29, 87
   preserved in translation, 40, 132
   properties of —, 31, 88
   See also valid inference.
consequence, syntactic
   predicate logic, 102
   propositional logic, 46
      properties of, 47
consequent, 8, 14
consistency of predications.
   See extensionality of predicates.
consistent theory, 48
context, 2, 37, 38, 41, 109
contradiction, 38.
   See also anti-tautology.
contradictory propositions, 8, 196
contraposition, 38
contrary propositions, 196
conventions on formalizing, 41
   "anyone" and "anybody", 124
   Converting Nouns to Predicates, 113
   Formal Theories, 117
   Grammar, 41, 114
   Negations, 114
   "No" and "Nothing", 147
   Parity of Form, 41, 114–115, 124, 129
   Parts to whole, 111
   Passive into Active, 126
   summary of —, 132–134
   There are exactly n, 146
   There are at least n, 145
   There are at most n, 145
   There are n things, 146
criteria of formalization, 40–41
   Categorematic Words, 116
   Metaphysics, 108
   Possibilities, 108
   summary of —, 132
cut rule, 47

De Morgan's laws, 33
decidability of tautologies,
   not for predicate logic, 104
   propositional, 26–28
deducibility, 46

Deduction Theorem,
  semantic, 31, 47
  syntactic, 47, 170, 175
definition by induction, 16
denying the antecedent, 11
derivation, 44
description and pointing, 166–168
direct way of reasoning with conditionals, 10.
  See also modus ponens.
disjunct, 14
disjunction, 14
  associativity and commutativity of, 33
  inclusive vs. exclusive, 22, 39
  law of addition, 33
  tautology principle, 33
distinguishability of things, 52, 76,
  86–87, 138.
  See also identity.
distribution axiom, 102
distribution laws, propositional, 33
distribution of quantifiers, 93–96, 102
Division of Form and Content,
  predicate logic, 82, 91
  propositional logic, 23
double negation, law of, 33

element. *See* things.
entity. *See* things.
equality predicate, 139–140
  interpretation as implicit identity, 140
  syncategorematic, 140, 143
equivalence relation, 143, 180–181
equivalent propositions, 9, 34, 71
"everyone", "everybody", 124, 134
excluded middle, 29
excluding possibilities, 7–8, 30
exclusive "or", 22, 39
existence,
  — is timeless, 122
  pointing and —, 168–169
existential assumption,
  "all" has —?, 82, 109, 110, 123
  categorical propositions, 196–197
  universe of a model and —, 77
existential generalization, 100, 177
existential quantifier, 58
  as disjunction, 155
explicit identity, 142–143

exportation, 33
extension of a predicate, 186–188
extensionality axiom. *See* substitution
  of equals.
extensionality of predicates, 72–73, 81, 183, 188

faith, 166
falsifying a conditional, method of, 27–28
finitely many things in universe, 155, 183
"follows from", 4
form, logical, 37
Form and Content, Division of.
  See Division of Form and Content.
Form and Meaningfulness,
  predicate logic, 66
  propositional logic, 24
form, propositional, 90
formal language,
  defined without induction, 18–19
  predicate logic, 61–62
  propositional, 16
formal logic, 13
  independent of subject matter, 76
formalization, 116
  analysis vs. —, 115–116, 117, 123
  anti-tautology preserved in —, 36, 40
  as translation, 116, 131
  context and, 37
  conventions for —. *See* conventions
    on formalization.
  criteria for —. *See* criteria of
    formalization.
formalizing a notion, 116–117
formula. *See* wff.
free variable. *See* variable, free.
Fully General Abstraction,
  classical predicate logic models, 189
  classical propositional logic models, 186

geometry, 103–104, 116
Gödel, Kurt, 104
grammar
  formalization preserves, 38, 41, 114, 116
  not sufficient guide for reasoning, 53, 110,
    128, 129, 130, 133, 192.
  See also parity of form; syntax.

habitual tense, 126–127

Hausdorff, Felix, 185
Heath, P. L., 147
hypotheses,
    formal proof from, 46
    reasoning from, 12, 31

identity, 138–139
    — axiom, 141
    implicit —, 138-143
    explicit —, 142–143
"if . . . then . . .". *See* conditional.
"if and only if ", 9,
"iff", 9
importation, 33
inclusive "or", 22, 39
indexical, 3
indirect way of reasoning with conditionals
    (*modus tollens*), 11
individual. *See* things.
induction, proof by, 16, 162–163
inductive definition, 15–16
inference, 4
infinite collections, 19, 173, 179, 190
infinitely many premises, 158
infinitely many things, 145
instantiation, 100, 102
interchange of premises, 33
interpretation, 85
"is" as identity, 140

Kleene, Stephen C., 124

length of a wff,
    predicate logic, 61–62
    propositional logic, 16
location as thing, 125, 130
logic. *See* formal logic.
logic and reasoning, 6, 13, 35, 45, 74,
    76, 87, 96, 103, 107, 135, 156–158,
    189–190, 191–192
logical form, 37
logical possibility, 156
logical vocabulary, 65

major premise, 198
major term, 198
masses and mass terms, 120–121,
    130–131, 135

material implication, paradoxes of, 33
Mates, Benson, 184
mathematical logic, 185–190
mathematics. *See* arithmetic; geometry;
    induction; mathematical logic.
meaning, 5
    formalized with axioms, 132
    preserved in translation, 131
    via formal theory, 134
meaning axioms, 115–117
mention of a piece of language, 3
metalanguage, 47, 183
metalogic, 76, 184, 186
metaphysics, 108, 132, 135
metavariables, 15, 20, 61, 79, 188
middle term, 198
minor premise, 198
minor term, 198
model(s)
    classical abstraction of —, 188
    classical mathematical —, 189
    — of Γ, 87
    predicate logic, 86
    propositional logic, 25–26
    sufficiently many, 26, 89
    type **I** vs. type **II**, 185–186, 188–190
modus ponens, 10, 43, 103
    generalized, 175
modus tollens, 11
"more", 153

name(s), 55–56, 165–169
    are types, 69
    Classical Abstraction of —, 69
    non-referring —, 123–124
    simple —, 63
    variables as temporary —, 57, 70, 182–184
name symbol, 61
naming, 69–71, 77, 165–169, 182–184.
    See also reference.
nand, 34
negation, 14
    double —, law of, 33
    formalizing, 39, 113–115
    predicate —, 113
    prefixes as —, 112–113
negative categorical proposition, 194
"neither . . . nor . . .", 8, 34

"no" (quantifier), 146–147
noncontradiction, law of, 29
nonempty universe, 77
"not", 14, 22–24. *See also* negation.
"nothing", 146–147
nouns into predicates. *See* conventions on formalizing.

object. *See* things.
objectivity, 5–6, 74
"only" (quantifier), 150–151
"only if ", 9, 38
open wff, 62
"or", 2–3
   inclusive vs. exclusive, 22
order relation, 117

paradoxes
   classical propositional logic, 33, 40
   self-reference, 75–76
parentheses, 14, 58,
   conventions on deleting, 17, 63
parity of form. *See* conventions on formalizing; grammar.
particular categorical proposition, 194
passive form of verbs, 125–126
attention, paying, 13, 88, 131, 158, 165–169
PC, 32
platonist, 5
   medieval logic and, 193
   on identity, 138–139
   on models, 186
   on predicates and predications, 73–74, 87
   on propositions, 5–6
   on reference, 74, 76, 87, 183–184
plural, formalizing, 109–110, 152
pointing, 57, 70, 76, 83, 87, 165–169
"possesses", 128–129
possibilities. *See* model(s); way the world could be.
predicate(s), 55–56
   abstract?, 73–74
   applies to object(s), 72, 74
   are concepts?, 74
   Aristotelian logic —, 191–192
   arity of, 56
   atomic?, 92
   binary, 56

predicate(s) (continued)
   classical abstraction of —, 186
   compound?, 92
   extension of a —, 187
   extensional, 73, 81, 183, 188
   holds of object(s), 72
   linguistic?, 73–74, 92, 189, 194
   list of all?, 61
   platonist conception. *See* platonist.
   predicated of object(s), 71
   satisfied by object(s), 72
   simple, 63
   true of object(s), 72
predicate logic, 59
   classical. *See* classical predicate logic.
   formal language, 61–62
predicate symbol, 61
predication, 71
   extensionality of, 73, 81, 183, 188
premise(s) of an inference, 4
   infinitely many?, 158
process, 56, 118, 121, 130, 135
pronouns, 57
proof by induction, 16
proof (formal), 44
proof from hypotheses, 12, 31, 46.
   See also reasoning from hypotheses.
proof sequence, 44
properties, 73–74, 131
proposition, 2–3
   Aristotelian classifications, 194–195
   as abstract object, 5–6, 74
   as closed wff, 65, 86
   as meaning, 5–6
   as thought, 5–6, 74
   as type, 2–3
   atomic. *See* atomic proposition.
   compound —, 7
propositional form of a wff in
   predicate logic, 90
propositional logic, 13
   classical, 32
   in predicate logic, 89–90
propositional variables, 14
provability and truth, 103–104
punctuation in a formal language, 56, 75.
   See also commas; semicolon.
pure reference, 86–87

qualities, 131
quality of a categorical proposition, 195
quantification, 58
    is timeless, 122–123
    relative, 108–113
quantifier(s), 58
    — binds a variable, 62,
    distribution of, 93–96, 102
    numerical, 147–148
    order of —, 93–94
    pair of, 93–94, 102
    permuting, 175
    referential interpretation, 183
    scope of —, 62
    substitutional interpretation —, 182–183
quantity of a categorical proposition, 195
questions, 1
quotation marks, 3
    eliminating, 14
    scare —, 3

Ralph. *See the back cover.*
range of variables, 77
realization,
    none in mathematical logic, 184
    predicate logic, 63–65
    propositional logic, 19
    substitutional interpretation, 182
reasoning. *See* logic and reasoning.
reasoning in a chain with conditionals, 11
reductio ad absurdum, 32, 96
reference, 69, 165–169
    can't talk about in semi-formal language, 75, 139
    platonist on —, 74
    pure, 86–87
    timeless, 122
referential interpretation of quantifiers, 183
relation, 56, 71, 72
relative adjective, 117
relative quantification, 108–113

"sake", 53
scare quotes, 3
scheme
    of axioms, 43, 46
    of tautologies, 26
    of valid inferences, 31

scope of quantifier, 62
self-reference, 75–76
Self-Reference Exclusion Principle, 75, 139
semantic consequence. *See* consequence, semantic.
Semantic Deduction Theorem.
    See Deduction Theorem.
semantics, 13
semicolon, 39
semi-formal language,
    predicate logic, 64
        none in mathematical logic, 184
        replaces ordinary language?, 107
    propositional, 19
sequence of objects, 74, 186–187, 189–190
set theory, 31, 185
Sheffer stroke, 34
Shoenfield, Joseph, 183–184
simple name, 63
simple predicate, 63
singular subject, 194
"some", 58, 83
    as disjunction, 155
    formalized as plural, 109–110, 152
    related to "all", 94, 103
    See also relative quantification.
"someone", "somebody", 124, 134
soundness of axiom system,
    predicate logic, 103,
    propositional logic, 74, 170
square of opposition, 196–197
standard form of a categorical proposition, 191
standard form of a syllogism, 198
strict implication, paradoxes of, 33
strong inference, 4
subcontrary propositions, 196
subformula, 91
subject matter, 21, 40, 45, 50
subject of a categorical proposition, 191
subject-predicate distinction, 51, 191
subject term in a categorical proposition, 191–192
substitution
    of equals, 141
    of variables, 97–100
    of wffs, 90–91
    uniform —, 90

substitutional interpretation of quantifiers, 182–183
sufficient collection of models,
   predicate logic, 89
   propositional logic, 26
superlatives, 150
syllogism(s), 198
   evaluating, 200–201
syncategorematic vocabulary, 65
   used in formalizing, 114–115
   See also categorematic vocabulary.
syntactic approach to logic, 43, 44–45, 53–54, 157, 192. *See also* grammar.
syntactic consequence. *See* consequence, syntactic.
syntax, 13. *See also* grammar.

tautology, 4
   preserved in formalizing, 40, 131
tautology, formal,
   predicate logic, 87
   propositional logic, 26
      decidability, 29–31, 45
      in predicate logic, 90
term(s), 61
   conjunction of —, 130
   free for a variable, 99
term, Aristotelian, 191
tertium non datur, 29
"the", 151–152
theorem (formal), 44
   used as axiom, 44
theory, 48
   complete, 48, 103
   consistent, 48, 103
   formal — of geometry, 116
   formal — of ordering, 117
things, 52, 159, 165–169
   all — ?, 76–77
   categorical propositions and, 192–193
   distinguishability of, 52
   predicate logic determines notion of, 52, 76–77, 139
Things, the World, and Propositions, 52–53, 131
thoughts as propositions, 5–6, 74
time in reasoning, 40, 50, 121–123, 129–130, 134, 183

transitivity of →, 33, 175
translation,
   formalizing as, 116, 131
   preserves meaning, 131
truth and provability, 103–104
truth function, 21, 23, 40
truth in a model,
   classical mathematical —, 189
   predicate logic, 86
   propositional logic, 25–26
truth-tables, 21–23
truth-value, 2
types,
   are abstract?, 6
   names are, 69
   propositions are, 2–3
   words are, 2–3

"un-" as prefix, formalizing, 114
unary predicate, 56
uniform substitution, 90
unique readability of wffs,
   predicate logic, 62
   propositional logic 16
uniqueness quantifier, 146
universal categorical proposition, 194
universal closure of a wff, 65
universal instantiation, 100
universal proposition (Aristotelian), 194
universal quantifier(s), 58
   as conjunction, 155
   pair of, 93–94, 102
universe for a realization, 76–77
use of a piece of language, 3
utterance, 2–3, 5

valid inference, 4, 156–160
   grammar determines?. *See* grammar.
   mathematical logic, 190
   predicate logic, 87
   propositional logic, 29
valid wff,
   in mathematical logic, 190
   predicate logic, 87
   propositional logic, 26
   See also tautology.
valuation predicate logic (closed wffs), 86
valuation propositional logic, 25

valuation relative to an assignment of
  references, 81
    extended to all wffs, 85
variable(s), 57, 182–184
  alphabetical order of, 65
  as placeholders, 182, 184
  as temporary names, 57, 70, 105, 182, 184, 193
  bound, 62
  free, 59, 61–62
  meta—. *See* metavariable.
  propositional —, 14
  range of —, 77
  substitution of —, 97–100
  term is free for a —, 99
verbs don't denote things, 118. *See also* process.

water, 120
way the world could be (possibility), 4, 25, 26, 50, 81, 88, 108, 116, 131, 156–160. *See also* model.
weak inference, 4
well-formed formula. *See* wff(s).

wff(s),
  atomic,
    predicate logic, 62
    predicate logic with equality, 141
    propositional logic, 16
  closed. *See* closed wff.
  collection of, 185
  compound
    predicate logic, 62
    propositional logic, 16
  length of,
    predicate —, 61–62
    propositional —, 16
  open, 62,
  propositional, 16
  predicate, 61
  subformula, 90–91
  substitution of —, 90–91
  unique reading of,
    predicate —, 62
    propositional —, 16
  See also formal language.
"whatever", 130
words as types, 2–3

Richard L. Epstein received his B.A. *summa cum laude* from the University of Pennsylvania and his Ph.D. from the University of California, Berkeley. He was a Fulbright Fellow to Brazil and a National Academy of Sciences scholar to Poland. He is the author of books on formal logic, informal logic, mathematics, and philosophy. He is now head of the Advanced Reasoning Forum in Socorro, New Mexico.

The Advanced Reasoning Forum is dedicated to advancing the study and teaching of formal logic, critical thinking, philosophy of language, linguistics, and ethics.

Other books by Richard L. Epstein now published by ARF
> *Critical Thinking*
> *Propositional Logics*
> *Predicate Logic*
> *Computability* with Walter Carnielli
> *Reasoning in Science and Mathematics*
> *Reasoning and Formal Logic*
> *The Internal Structure of Predicates and Names*
> *Conventional Gestures: Meaning and Methodology*

www.AdvancedReasoningForum.org

www.ingramcontent.com/pod-product-compliance
Lightning Source LLC
Chambersburg PA
CBHW082118230426
43671CB00015B/2729